PENGUIN BOOKS

NUKESPEAK

Stephen Hilgartner writes on issues concerning technology and public policy. Richard C. Bell is a journalist and political organizer interested in environmental and public-policy issues. Rory O'Connor is a print and broadcast journalist and the author of articles for *The Atlantic*, *Newsweek*, *Mother Jones*, *Rolling Stone*, and other publications. All three authors live in the Boston area.

NUKESPEAK

*Stephen Hilgartner, Richard C. Bell,
Rory O'Connor*

PENGUIN BOOKS

Penguin Books Ltd, Harmondsworth,
Middlesex, England
Penguin Books, 40 West 23rd Street,
New York, New York 10010, U.S.A.
Penguin Books Australia Ltd, Ringwood,
Victoria, Australia
Penguin Books Canada Limited, 2801 John Street,
Markham, Ontario, Canada L3R 1B4
Penguin Books (N.Z.) Ltd, 182–190 Wairau Road,
Auckland 10, New Zealand

First published in the United States of America by
Sierra Club Books 1982
Published in Penguin Books 1983

LIBRARY OF CONGRESS CATALOGING IN PUBLICATION DATA
Hilgartner, Stephen.
Nukespeak.
Includes bibliographical references and index.
1. Atomic energy—United States. 2. Atomic
energy—United States—Terminology. I. Bell,
Richard (Richard C.). II. O'Connor, Rory.
III. Title.
QC773.3.U5H54   1983   363.1′79   82-22268
ISBN 0 14 00.6684 5

Printed in the United States of America by
Fairfield Graphics, Fairfield, Pennsylvania

To George Orwell

"A change in language can transform our appreciation of the Cosmos."
—*Benjamin Lee Whorf*

Table of Contents

List of Figures and Plates

PLATES

Acknowledgments

This book would not have been possible without the help of many people. We would like to thank the following people and organizations for providing documents, graphic materials, and their personal comments and recollections: Bob Alvarez of the Environmental Policy Center; the Atomic Industrial Forum; John Cronin of the *Boston Herald American*; Roger Anders, Tom Bauman, Marlene Flor, Louis Hicks, Murray Nash, and Frank Standerfer of the Department of Energy; Elaine Douglass; Mike Segal and Trey Taylor of the Edison Electric Institute; Dr. Rose Goldman; Harold Knapp; Ken Kronenberg; William Chambers, Robert Porten, and Faith Stevens of the Los Alamos Scientific Laboratory; Arlen J. Large of the *Wall Street Journal*; Sam Love; Dr. Joseph Lyon; Daniel Kleitman, Philip Morrison, and Norman Rasmussen of the Massachusetts Institute of Technology; Dr. Karl Morgan; Howard Morland; the National Archives; Peter Bradford, Raymond Brady, and Frank Ingram of the Nuclear Regulatory Commission; Kevin O'Neil; Carolyn Lewis and Anne Trunk of the President's Commission on the Accident at Three Mile Island; Edward Radford; Peggy Bennett and John Deichman of Rockwell Hanford Operations; Mitchel Rogovin; J. Newell Stannard; Robert Pollard and Lois Traube of the Union of Concerned Scientists; Paul Walker; and Ruth Young of the *Bulletin of the Atomic Scientists*. We would also like to thank all of Boston's late-night jazz radio hosts, especially Steve Elman of WBUR and Eric Jackson of WGBH, for playing the music that helped us write through the night.

The following people offered much appreciated advice and assistance: Helena Bentz; Sam Dorrance; John Gliedman; Kevin Gorman; David Jhirad; David Kreutzer; Winona LaDuke; Tom Lifson; Renee Parsons; and Ron Rosenbaum.

We are grateful for the thorough research assistance we received from: Linda Bailey; Lisa Barnett; Cliff Bob; Bob Brainerd; Terry Byrne; Ed Holston; Lisa Mosczynski; and Heidi Schmidt.

We are indebted to the following people for reading the manuscript and providing comments: Thomas Babe; Neal Bell; Peter Bradford; I.C. Bupp;

Joe Cone; Mary Filmore; Jim Hilgartner; Harold Knapp; Ronnie Lipschutz; Amory and Hunter Lovins; Philip Morrison; and Michael Reich. We found their criticism valuable. Any errors are our own. We would like to specially thank Dr. Benjamin Spock and Mary Morgan for their comments and their unflagging support.

We are grateful to Don Cleary for helping us develop the concept for the book, to Wendy Goldwyn of Sierra Club Books who initiated the project, to our editor, Daniel Moses, who guided us through the process of writing and editing, and to JoAnn Pluemer who copyedited the manuscript.

There are five people to whom we are particularly grateful for their contributions to the project, for their patience and support, and for always coming through in a pinch:

John Heymann produced crisp graphics from our muddled verbal descriptions. We thank him for his patience and for calmly meeting frantic deadlines.

We thank Debbie Hawkins for all of her help, and especially for tolerating the transformation of her home into a round-the-clock workspace. Richard Bell would like to thank his wife, Debbie, for the love and support that sustained him during this project.

We thank Alison Humes for her patience and assistance on both a professional and a personal level; her reaction to an early draft was very helpful.

Howard Shrobe of the Artificial Intelligence Laboratory at MIT was involved in every step of the project. We thank him for his insightful comments, for his editorial help, and for working closely with the authors on several chapters of the manuscript.

We thank Kathleen O'Neil for her dedication and for performing a myriad of tasks, including tracking down obscure documents and photographs, obtaining permissions, conducting research, and preparing footnotes. She was an integral participant in every phase of the project.

Introduction

The language we use has important influences on our thinking.

George Orwell wrote passionately about the role of language in shaping political discourse and understanding. In his 1946 essay, "Politics and the English Language," Orwell charged that political language consists mainly of "euphemism, question-begging and sheer cloudy vagueness." Thus "pacification" is the bombing of defenseless villages; "transfer of population" or "rectification of frontiers" is the robbing of millions of peasants of their farms; "elimination of unreliable elements" is accomplished when "people are imprisoned for years without a trial or shot in the back of the neck or sent to die of scurvy in Arctic lumber camps."

In the thirty-six years since the atomic bombings of Hiroshima and Nagasaki, a new language has evolved. We call that language *Nukespeak*. Nukespeak is the language of nuclear development, a term we use to include the development of both nuclear weapons and nuclear power. In Nukespeak, atrocities are rendered invisible by sterile words like *megadeaths*; nuclear war is called a *nuclear exchange*. Nuclear weapons accidents are called *broken arrows* and *bent spears*. Plutonium is called a *potential nuclear explosive*. The accident at Three Mile Island was called an *event*, an *incident*, an *abnormal evolution*, a *normal aberration*, and a *plant transient*. India called its nuclear bomb a *peaceful nuclear device*.

Nukespeak is the language of the nuclear mindset—the world view, or system of beliefs—of nuclear developers. The word *mindset* means what it implies, a mind that is already set. A mindset acts like a filter, sorting information and perceptions, allowing some to be processed and some to be ignored, consciously or unconsciously. Nukespeak encodes the beliefs and assumptions of the nuclear mindset; the language and the mindset continuously reinforce each other.

Euphoric visions of nuclear technologies are an important expression of the nuclear mindset. The discovery of X-rays and radium at the end of the nineteenth century brought forth visions of a technological *Garden of Eden*. The *philosopher's stone* and the *elixir of life* had been found at last. The discovery of nuclear fission in 1938 unleashed a torrent of similar imagery: nuclear-powered planes and automobiles would whisk us effortlessly around the globe, while unlimited nuclear electricity powered underground cities, farms, and factories.

After World War II and the bombing of Hiroshima and Nagasaki, nuclear developers used information-management techniques—official secrecy and public relations—to promote what one called the *"sunny side of the atom."* Once atomic energy was applied for peaceful purposes, a nuclear-powered *paradise* would be at hand. Electricity *"too cheap to meter"* would power the new *Golden Age.* No problem would be too difficult to solve quickly and economically. Reactors would operate safely; the disposal of radioactive waste would be a nonproblem; international *safeguards* would allow the benefits of the *peaceful atom* to spread to every nation while preventing the spread of nuclear bombs.

In his novel *1984,* Orwell warned about the danger of complete state control over the dissemination of information. The history of nuclear development has been profoundly shaped by the manipulation of information through official secrecy and extensive public-relations campaigns.

Nukespeak and the use of information-management techniques have consistently distorted the debate over nuclear weapons and nuclear power. Time and time again, nuclear developers have confused their hopes with reality, publicly presented their expectations and assumptions as facts, covered up damaging information, harassed and fired scientists who disagreed with established policy, refused to recognize the existence of problems, called their critics mentally ill, generated false or misleading statistics to bolster their assertions, failed to learn from their mistakes, and claimed that there was no choice but to follow their policies.

The nuclear debate is the most important debate in human history. We hope that this book will help readers understand the language, visions, and mindset underlying nuclear development, and that this understanding will contribute to a fruitful redirection of the discussion.

NOTE TO THE READER

Throughout the text, we have highlighted Nukespeak words and phrases with italics. We have indicated wherever italics within a quotation are due to the author's original emphasis. In all other cases, italics were added. We have included a special index of Nukespeak terms.

PART ONE

Nuclear Visions

Our world faces a crisis as yet unperceived by those possessing the power to make great decisions for good or evil. The unleashed power of the atom has changed everything save our modes of thinking, and thus we drift toward unparalleled catastrophe.[1]

Albert Einstein
May 23, 1946

CHAPTER ONE

The Elixir of Life

The invisible light of radiation burst onto front pages around the world on January 6, 1896, with the announcement of the discovery of mysterious new rays by the German scientist Wilhelm Conrad Roentgen. Although some scientists called them Roentgen Rays, Roentgen himself preferred a different term, X-rays—X standing for unknown. The equipment for producing X-rays was widely available, and scientists rushed to explore the new phenomenon.

The uses of X-rays proliferated so fast that only a month after the discovery was announced in the United States, the *New York Times* editorialized: "Professor Roentgen's discovery has been followed by a crop of new words large enough almost to warrant new editions of all the dictionaries. Already it is necessary to puzzle over the meaning of 'skiagraphy,' 'cathodograph' [both terms for X-ray photographs], and a dozen other terms, some of them freshly minted and the rest employed in wholly unfamiliar significations. . . ."[2]

Within four months, X-rays had been used to test welds, to locate bullets in gunshot wounds, to set broken bones, to kill diphtheria microbes, to diagnose tuberculosis, pneumonia, and enlarged hearts and spleens, and to treat cancer. X-ray exhibits and demonstrations became popular events at fairs across the country. Thomas Edison, who moved so quickly to explore the new technology that many Americans believed he discovered X-rays, mounted the first major public demonstration in the United States at the exhibition of the National Electric Light Association in New York City in May, 1896. Crowds lined up for a chance to "see their bones."[3]

By 1898, personal X-rays had become a popular status symbol in New York. The *New York Times* reported that "there is quite as much difference in the appearance of the hand of a washerwoman and the hand of a fine lady in an X-ray picture as in reality. . . ." The hit of the exhibition season was Dr. W. J. Morton's full length portrait of "*the X-ray lady*," a "fashionable woman who had evidently a scientific desire to see her bones." The portrait was said to be a "fascinating and coquettish" picture, the lady having agreed to be photographed without her stays and corset, the better to satisfy the "longing to have a portrait of well-developed ribs." Dr. Morton said women were not afraid of X-rays: "After being assured that there is *no danger* they take the rays without fear."[4]

The titillating possibility of using X-rays to see through clothing or to invade the privacy of locked rooms was a familiar theme in popular discussions of X-rays and in cartoons and jokes. Newspapers carried advertisements for *"X-ray proof underclothing"* for those seeking to protect themselves from X-ray inspection.[5] An 1896 photography magazine carried the following poem, which captured this mood:

X-actly So!

The Roentgen Rays, the Roentgen Rays,
What is this craze?
The town's ablaze
With the new phase
Of X-rays's ways
I'm full of daze
Shock and amaze;
For nowadays,
I hear they'll gaze
Thro' cloak and gown—and even stays
These naughty, naughty Roentgen Rays[6]

Some of the proposed uses of X-rays exceeded the imagination of even the wittiest of humorists. For example, in 1911, experimenters at the University of Pennsylvania made plans to use X-rays to take a photograph of the human soul. The *New York Times* consulted Dr. Duncan MacDougall on the question of whether this attempt might succeed. (MacDougal was an expert on the weight of the soul, which he estimated at between ½ and 1½ ounces.) According to MacDougall, "at the moment of death the soul substance might become so agitated as to reduce the obstruction that the bone of the skull offers ordinarily to the Roentgen ray and might therefore be shown on the plate as a lighter spot on the dark shadow on the bone."[7]

Some people thought that X-rays might be useful for saving energy. In a 1931 speech to the Boston Chamber of Commerce, Dr. Willys R. Whitney, director of research for the General Electric Company, outlined experiments he had conducted on staying warm with X-rays. According to Whitney, "It might be useless to heat our houses, with all their contents, including the air, if we would get along by internally heating ourselves [with X-rays]."[8]

Whitney's proposal was based on observations that men's blood temperature had been seen to rise in proximity to X-ray apparatus, and that fruit flies had been revived from seeming hibernation or death in frigid containers by the application of X-rays.

Whitney reported that he and his associates had conducted an experiment on an X-ray-heated rat cage. The rats were enclosed in a long ice-cooled glass box; at one end, the rats could warm themselves in an X-ray field. The rats clearly preferred the field end, and every day the experimenters increased the intensity of the field.

The experiment came to an abrupt end, Dr. Whitney said: "No one knows how far we might have gone, but one day one of the rats came hurriedly out of his warm bed, leaving his tail in the cotton. The tail proved to be entirely dried out. The rat was *unhurt* except in appearance, and ate out of my hand at once."[9]

X-rays were not the only new force announced in 1896. Later that year, the French scientist Henri Becquerel found other mysterious rays coming from uranium ores, though they were much weaker than X-rays. Following up on Becquerel's discovery, Pierre and Marie Curie announced their discovery of a new chemical element, radium, which gave off a variety of intense rays. This new phenomenon was called radioactivity. Today we know that the Curies were witnessing the spontaneous disintegration of atomic nuclei, which were giving off alpha and beta particles and gamma rays.

Sir William Ramsay, one of the leading experts in the new field of radioactive substances, thought there were no limits to what radium might mean to the world. He wrote that the *"philosopher's stone* will have been discovered, and it is not beyond the bounds of possibility that it may lead to that other goal of the philosophers of the Dark Age—the *elixir vitae.*"[10]

Frederick Soddy was also enraptured by the properties of radium. Soddy, a popular author and public lecturer, was one of the pioneers in the physics and chemisty of radioactive substances, winning the Nobel Prize for chemistry in 1921. In his widely read book, *The Interpretation of Radium and the Structure of the Atom*, Soddy wrote about the radioactive properties of radium:

> There is something sublime about its aloofness from and its indifference to its external environment. It seems to claim lineage with the worlds beyond us, fed with the same inexhaustible fires, urged by the same uncontrollable mechanism which keeps the great suns alight in the heavens over endless periods of time.[11]

The luminous properties of radium soon produced a full-fledged radium craze. A famous woman dancer performed *radium dances* using veils dipped in fluorescent salts containing radium. According to one recent history of radioactivity in America during this period, *radium roulette* was popular at New York casinos, featuring a "roulette wheel . . . washed with a radium solution, such that it glowed brightly in the darkness . . . an unseen hand cast the ball on the turning wheel and sparks marked its course as it bounded from pocket to glimmery pocket."[12] A patent was issued for a process for making women's gowns luminous with radium, and Broadway producer Florenz Ziegfeld snapped up the rights for his stage extravaganzas.[13]

Dr. W. J. Morton, the successful promoter of *"the X-ray lady,"* was also an enthusiastic booster of the medicinal use of radium to produce what he called *"Liquid Sunshine."* Doctors had already been using external exposure to visible light to treat various skin diseases. Morton showed that solutions of certain substances, such as an extract made from the inner skin of the horse chestnut, would fluoresce—or give off light—when exposed to radium. He argued that if such a solution were taken internally, and then a radium source were placed next to the skin, the rays would produce beneficial fluorescence within the body, healing diseased organs.

Dr. Morton introduced Liquid Sunshine at a January 1904 meeting of the Technology Club, the New York City alumni association of the Massachusetts Institute of Technology. He claimed that with Liquid Sunshine, "it will be possible to bathe a patient's entire interior in violet or ultraviolet light . . . and this light we have decided to call 'sunshine.' We know of the value of sunshine on the outside . . . and we believe it will have a similar effect on the inside."[14]

At their annual meeting in February, 1904, the Technology Club showed its enthusiasm for Dr. Morton's new proposal by mounting a "Liquid Sunshine Dinner." Everyone was equipped with a small glass of the fluorescent Liquid Sunshine. When the lights were dimmed for viewing the glowing cocktails in all their splendor, the entertainment chairman appeared

> with a phosphorescent cigar in his mouth which looked like a gigantic icicle . . . human skeletons, covered in luminous paint emerged . . . and proceeded to dance. Minute balloons, similarly treated, . . . floated here and there. . . . Two pasteboard chickens decorated with the paint were shown fighting over an egg. . . .[15]

At the climax of this ghostly dinner, the guests toasted "Tech's welfare and future" by downing their Liquid Sunshine "amid great applause and enthusiasm."[16] The *New York Times* reported that "[n]one of the self-sacrificing scientists who drank the liquid became transparent afterwards, although all were assured that, as a matter of fact, their interiors were thoroughly illuminated. . . ."[17] The MIT alumni bulletin reported that after this rousing toast, "every fellow went home feeling that the long-wished-for 'Tech Spirit' was at last awakened in New York City."[18] (Not all doctors were as impressed as the Technology Club by Dr. Morton's Liquid Sunshine. Some physicians said that a more appropriate name would be "liquid moonshine.")[19]

There was a widespread belief that low doses of radium were beneficial, stimulating the life processes. This hypothesis was supported by the discovery that radon, a radioactive element produced by the decay of radium, was present in the air and water at many hot springs and spas. Might this radioactivity explain why a hot spring could be healing, while a hot bath at home was not?

The belief in the life-giving properties of low doses went far beyond such crude analogies. Professor John Burke of the world-famous Cavendish Laboratories at Cambridge University thought he had produced a new form of life, which he called "*radiobes*," by sealing up a radium source in tubes of sterilized bouillon.[20] And Doctor H. H. Rusby, dean of the Columbia University College of Pharmacy and federal pharmacologist for the Port of New York, claimed that his experimental studies showed that radium was a wonderful fertilizer. Not only were crop yields raised, reported Rusby, but the vegetables tasted better as well, "a matter of special comment by all who tried them." As far as costs to farmers were concerned, Rusby said that the increased market value of even the first year's crops would more than pay for the *radioactive fertilizer.*[21]

Some of the most outlandish claims for the utility of radium as a life-enhancing force came from the medical community, a reflection in part of the turn-of-the-century physician's limited armamentarium, especially for the treatment of cancer. Radium appeared to be the magic bullet that could replace the crude and bloody interventions of the surgeon against cancer. Dr. Howard A. Kelly, one of the most prominent promoters of radium, wrote in 1913: "Radium is destined to produce a change in surgical and medical work not less marked than the introduction of the Roentgen ray, perhaps even more decided. Radium in surgery will definitely cure many forms of cancer. . . ."[22] Another radium enthusiast found in 1916 that the physiological action of radium was

> . . . not unlike a fairy tale. It often increases the red blood count. . . . It stimulates all cell life, particularly that of the enzymes thus aiding and improving metabolism. It increases the elimination of carbon dioxide, urea, and uric acid. . . . It stimulates and increases the appetite, and aids digestion by activating the digestive ferments. . . . It lowers blood pressure almost so surely as you administer it. . . . No toxic or lasting ill effects have been reported.[23]

The medical profession was particularly enthusiastic in its use of radium and X-rays on women, to treat both cancer and a host of non-life-threatening ailments. According to *Radiologic Maxims*, a pithy little 1932 book approved by leading medical professors, "Radium therapy in gynecology is essentially the handmaiden of the surgeon."[24]

Radiation was thought to be useful in inducing menopause artificially. "The menopause brought on by X-ray or radium treatment is less stormy than that produced by castration."[25] "The only constant symptom of the radiotherapeutic menopause is the so-called 'hotflash.' "[26] Radiation was also said to be helpful to women having trouble with their natural menopause. "Most brilliant and satisfactory results are frequently obtained by radiation therapy in the treatment of women undergoing the stress and storms of the menopause, with mental, nervous, and physical break-downs."[27]

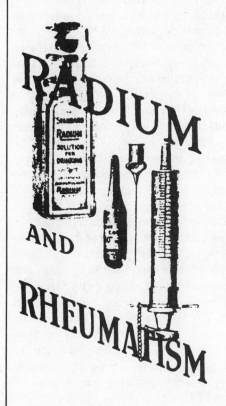

WHAT RADIUM DOES--INTERNALLY

1. Produces a general stimulation.

2. Improves blood picture.

3. Influences abnormal blood pressure.

4. Increases urine secretion.

5. Has a decided antiphlogistic and antineuralgic action.

6. Exerts a selective action in arthritis.

"Standard" Radium Solutions are permanently Radio-active. Radium Content Standardized and Guaranteed.

Send for Clinical and Descriptive Literature to

RADIUM CHEMICAL CO.
PITTSBURGH, PA.

FIGURE 1 The wondrous properties of radium were widely touted in advertisements like this one from the professional journal *Radium*.

Radiation was also thought useful for regulating menstrual functions. Dr. Thomas H. Cherry, professor of gynecology at the New York Post-

Graduate Medical School and Hospital and author of *Gynecologic Technic*, told readers of *Radiologic Maxims* that women with menorrhagia (excessive menstrual flow) could be treated by inserting tubes containing radium through the vagina into the uterus. He claimed this procedure would not damage the ovaries, and that some patients might miss a few periods before they "gradually assume a reduced and more normal flow."[28]

Lest physicians be worried about effects on the sex lives of women treated with radiation, the *Maxims* noted that "when menstruation is arrested by X-ray or radium, sexual desire and response are lessened or lost in only about twenty per cent of cases."[29]

We should note that these promoters of the use of X-rays and radium whom we have quoted were not *radiomaniacs* or *radiografters*, to use terms employed at a 1908 meeting of the American Roentgen Ray Society to describe men who were even more extravagant users of X-rays.[30] The men we have mentioned were among the foremost physicians, physicists, biologists, and chemists of their time, holding high-ranking medical-school and research positions: they were undisputed *experts*.

Even while the unrestrained use of X-rays and radium was growing, evidence was accumulating that the new forces might not be so benign after all. Hailed as tools for fighting cancer, they could also cause cancer. Doctors using X-rays were the first to learn this bitter lesson. By 1905, newspaper stories about the latest "martyr" to science were appearing with grisly regularity, as yet another X-ray experimenter died or suffered amputations to stave off X-ray-induced cancer. Much of this disease was brought on by physicians using their hands to calibrate X-ray equipment. X-ray tubes in the early years were highly variable in output, and physicians often used the density of the shadows cast by their hands on the fluoroscope screen to estimate the output of a tube. They also used their hands to demonstrate the apparatus to skeptical or fearful patients.[31]

The suffering of these early experimenters from *burns* or *X-ray dermatitis* or *X-ray eczema* was immense. As Dr. John Hall Edwards said, "I have not experienced a moment's freedom from pain for two years. In cold weather I am unable to dress myself, and the pain experienced cannot be expressed in words. Drugs have so far failed to give me the slightest relief. . . . The pain is of a neuralgic character, it never ceases and is from time to time intensified by sudden stabs and jumps of such severity as to make one cry out."[32]

In retrospect, it is easy to see that most of these deaths could have been avoided, if users had heeded the early warning signs. It was, after all, quickly apparent that these new forces had a marked impact on living tissue. Reports of painful and prolonged irritation of the skin began to appear in the scientific literature within months of Roentgen's announcement. Nor was this information buried in obscure medical journals. In a July 31, 1897 editorial, the *New York Times* warned of "Roentgen Ray Dangers": "Cases of severe burns are many and notorious. . . . Its destructive action has been discussed by high electrical authorities and the press has published their views. Everybody ought to know by this time that the Roentgen ray apparatus . . . is liable to produce serious injury. . . ."[33]

But safety costs money, and there were no regulatory agencies around to insist that manufacturers offer safer equipment, or even to provide a forum in which the claims of those who said there was *no danger* could be questioned. William Rollins, an experimenter who recognized the dangers of X-rays (which he called X-light) as early as 1898, built and wrote about equipment that would reduce the risk. But in 1903, Rollins observed that his ideas had not been widely adopted:

> Most of these precautions are neglected even at the present time, as may be seen by examining the illustrations in the catalogues of makers of apparatus, and in the papers and books of those who are writing on the subject, where open [unshielded] tubes are almost invariably figured. . . . That inefficient means are still employed . . . is partly due to attempts to ignore or disparage the crucial experiments that have been reported in other notes on the effects of X-light on animals. . . .[34]

Equipment manufacturers were not alone in playing down the dangers of X-rays. Dr. Mihran Kassabian, one of the foremost practitioners of the new science in the United States, did not even want physicians to use the term *burns* to describe the skin changes caused by overexposure to X-rays. (The technique of eliminating objectionable words is characteristic of Nukespeak.) According to the American Roentgen Ray Society historian Percy Brown, "Kassabian was worried that if 'burns' came to be associated, in the popular mind, with the medical use of the Roentgen rays, the progress of a new science might be definitely inhibited at a crucial period in its development."[35]

THE RADIUM DIAL WOMEN

The case of the *radium dial women* is a classic example of the problems that radiation victims encounter in trying to prove that radiation exposure has damaged their health. In 1925, front-page stories appeared alleging that a group of former employees at an East Orange, New Jersey plant of the United States Radium Corporation were dying from "radium poisoning." The women had all painted numbers on watches using a radioluminescent paint called *Undark*, which contained radium. In applying the paint, the women had used their lips to point their brushes, thus swallowing quantities of radium every day for several years. The case culminated in a lawsuit for damages in 1928.[36]

The women had all been forced to leave work by a variety of afflictions, which were initially diagnosed as mouth sepsis, anemia, necrosis of the jaw, syphilis, and Vincent's angina. It was well established by the mid-1920s that external sources of radiation—sources outside the body, like X-rays—could cause burns and cancer. It was not well understood that people could also be exposed to radiation from radioactive materials incorporated into the body. Once inside the body, radioactive materials function as internal emitters, bombarding nearby tissues with an ongoing dose of radiation. In contrast, exposure from external sources ceases as soon as the person and the radiation source are separated. The failure to distinguish between internal and external exposure remains one of the most common misunderstandings in public discussions of the effects of radiation (see chapter 7).

Radium was first identified as a potentially dangerous internal emitter by a New York dentist, who said that one of his patients who had worked at the radium plant was suffering from what he called "*radium jaw*."[37] This report interested an insurance company statistician, Dr. Fredrick L. Hoffman, who then gathered statistics on workers in Orange and other places who had died of "anemia." On the basis of this epidemiological work, Hoffman concluded that the unusual number of cases of "anemia" among U.S. Radium's employees was due to radium ingestion from tipping the brushes. He renamed the disease *radium necrosis*.[38]

The Orange County medical examiner soon found that radium, which is chemically similar to calcium, is selectively concentrated in the bones. Radioactive elements with this property are now called *bone-seekers*. Since blood is formed in the bone tissues, the radium caused various bone diseases and bone cancers, and damaged the blood-forming tissues as well, producing the "anemias."

The company denied that radium was responsible for the women's illnesses, laying out many of the lines of defense that nuclear developers have been using ever since in cases of possible radiation-induced disease. U.S. Radium first cited the small quantities involved, referring to the then generally accepted belief that small amounts of radium were stimulating. "Only in its intensive form is radium destructive. . . . In infinitesimal quantities it is commonly prescribed and is frequently given internally." If radium "exerted any effect at all," the company argued, it would be "a beneficial and not a baneful effect."[39]

The company also tried to blame the victims, claiming that the women had disobeyed instructions when they used their lips to point the brushes. In court, however, the former president of U.S. Radium demolished this defense, testifying that he did not warn the women of any danger involved in handling the paint.[40]

U.S. Radium also said that it had hired many women who were "unfit" for strenuous labor and were "already in comparatively poor health." Therefore the women were mistaken in blaming Undark when they "declined normally."[41]

Finally, the company suggested that the women's problems were psychological, not physical. "Radium, because of the mystery which surrounds much of its actions, is a topic which *stimulates the imagination*, and to our mind, it is to this and not to *actual fact* that many of the reports of the luminous paint's effects in our plant may be attributed."[42]

As the women's suit against U.S. Radium moved toward trial, an argument arose over the statute of limitations, which was two years. The company argued that even if radium had caused the afflictions, the company was not liable because the statute of limitations had expired. It is now well established that a considerable period of time—the latency period—may elapse between an exposure to radiation and the subsequent appearance of a disease such as cancer. Latency periods may be twenty years or longer, making it difficult in many cases to establish a cause-and-effect relationship between a cancer and previous radiation exposure.

In the case of the radium dial women, the judge ruled that since the women's bones still contained detectable concentrations of radium, "the statute of limitations does not apply until the period of injury ends, and these girls are still being injured."[43]

U.S. Radium settled out of court a few days before the official opening of the trial. The settlement included $10,000 cash for each woman, a $6000-per-year pension, medical treatment, and the payment of lawyers' fees.[44] Nominal sums were paid to the women's husbands for the "loss of their wives' services."[45] The company said it agreed to the settlement because of *"humanitarian considerations,"* refusing to admit, despite settling out of court, that radium was the culprit.[46]

CHAPTER TWO

The World Set Free

After the discovery of radioactivity, scientists quickly realized that there were much more powerful forces than radioactivity waiting to be freed from the atom. Could humanity learn to release these forces? If these forces were released, could society learn to control them? And, if so, at what cost?

In his 1909 book *The Interpretation of Radium and the Structure of the Atom,* Frederick Soddy predicted that scientists would eventually release vast amounts of energy from uranium. In a comparison remarkably similar to those made by nuclear promoters decades later, Soddy claimed that "the energy in a ton of uranium would be sufficient to light London for a year. The store of energy in uranium would be worth a thousand times as much as the uranium itself, if only it were under our control and could be harnessed to do the world's work in the same way as the energy in coal has been harnessed and controlled."[1]

At the time the possibility of harnessing uranium seemed remote, but Soddy was convinced that the day would eventually arrive:

> Looking backwards at the great things science has already accomplished, and at the steady growth in power and fruitfulness of scientific method, it can scarcely be doubted that one day we shall come to break down and build up elements in the laboratory as we now break down and build up compounds, and the pulses of the world will then throb with a new source of strength as immeasurably removed from any we at present control as they in turn are from the natural resources of the human savage.[2]

Soddy's book had a powerful impact on the well-known writer H. G. Wells. Wells realized, unlike Soddy, that the power of the atom might be used for evil as well as for good. In 1914, Wells published *The World Set Free,* an astoundingly prescient novel about the possible dangers and benefits that releasing atomic energy could produce. Wells dedicated the book to Soddy's *Interpretation of Radium.*[3]

Wells predicted that in the 1930s, scientists would release atomic energy, and that by the 1950s, this knowledge, in the form of what he called "*atomic bombs,*" would lead to a devastating worldwide nuclear war. This "*last war*" would leave hundreds of the world's cities in ruins. In *The World Set Free*, the exploding atomic bombs produce "puffs of luminous radio-active vapour drifting sometimes scores of miles from the bomb center and killing and scorching all they overtook."[4] The bombings set up firestorms like those at Hiroshima and Nagasaki. Victims of these attacks suffer from radiation burns that were "very difficult to heal."

In his novel, Wells poses the question: why were people blind to the danger of atomic energy? A fictional historian describes the period before the war as one that "'believed in established words and was invincibly blind to the obvious in things.'" The narrator continues:

> Certainly it seems now that nothing could have been more obvious to the people of the earlier twentieth century than the rapidity with which war was becoming impossible. And as certainly they did not see it. They did not see it until the atomic bombs burst in their fumbling hands. Yet the broad facts must have glared upon any intelligent mind. All through the nineteenth and twentieth centuries the amount of energy that men were able to command was continually increasing. Applied to warfare that meant that the power to inflict a blow, the power to destroy, was continually increasing. There was no increase whatever in the ability to escape.[5]

Wells understood the moral dilemmas and the guilt that scientists working on atomic physics would face. The fictional scientist Holsten, on the day after achieving the atomic breakthrough, walks disconsolately through London, feeling that he is "something strange and inhuman, a loose wanderer from the flock returning with evil gifts from his sustained unnatural excursions amidst the darknesses and phosphorescences beneath the fair surfaces of life."[6]

Is it possible for a scientist to hide his discoveries in order to protect the world? Holsten briefly considers not publishing his work, handing it over instead to "some secret association of wise men" who would protect it "until the world was riper for its practical application."[7] But scientific secrets are hard to keep, since the general principles on which new discoveries are based are widely known. In the end, Holsten decides to release his discovery, believing that even if he were to burn his papers, "before a score of years had passed some other man would be doing this. . . ."[8]

Holsten's discovery leads to atomic bombs and war. Can any good come from such an overwhelming demonstration of the danger of atomic energy? Wells takes a position similar to that taken by many people after Hiroshima and Nagasaki: here at last is a weapon so destructive that the peoples of the world will finally set up an international government to control atomic energy and outlaw war. The first step toward peace is to round up every atom of the fictional explosive element "Carolinum."[9] Then atomic energy can be used to create an age of peace and prosperity over the entire planet:

> Men spread now, with the whole power of the race to aid them, into every available region of the earth. Their cities are no longer tethered to running

water and the proximity of cultivation, their plans are no longer affected by strategic considerations or thoughts of social insecurity. The aeroplane and the nearly costless mobile car have abolished trade routes. . . . [Their cities] lie out in the former deserts, these long wasted sunbaths of the race, they tower amidst eternal snows, they hide in remote islands and bask on broad lagoons. . . .Every year the work of our scientific laboratories increases the productivity and simplifies the labour of those who work upon the soil, and the food now of the whole world is produced by less than one per cent of its population, a percentage which still tends to decrease.[10]

The World Set Free had a profound effect on Leo Szilard, a Hungarian-born physicist who played a pivotal role in the creation of the *Manhattan Project*, the atom-bomb-building program undertaken by the United States during World War II. Like Wells's fictional scientist Holsten, Szilard made the conceptual breakthrough that would eventually lead to nuclear weapons. And like Holsten, Szilard immediately thought of trying to keep his discovery secret.

From 1923 to 1933, Szilard lived in Berlin, dividing his time between the Kaiser Wilhelm Institute and the University of Berlin. He was closely associated with Albert Einstein at the University of Berlin, and the two men shared seven German patents on pumps and refrigeration systems, including a noiseless household refrigerator that used a liquid metal pumping system. Szilard first read *The World Set Free* in 1932, while he was still in Berlin.[11]

Szilard fled from the Nazis in 1933, after the Reichstag fire and Hitler's rise to power. He went to England and continued to study physics. His dangerous discovery was made in reaction to a speech he read given by Nobel-prize-winning physicist Ernest Rutherford. At a meeting of the British Association for the Advancement of Science in September, 1933, Rutherford warned those looking to liberate atomic energy that their expectations were "the merest moonshine."[12] This categorical denial engaged Szilard's imagination; as he said later, "Pronouncements of *experts* to the effect that something cannot be done have always irritated me."[13]

Szilard kept up with the latest work in nuclear physics, and he was familiar with James Chadwick's discovery of the neutron in 1932. Neutrons are electrically neutral particles that, together with positively charged protons, make up the nuclei of all atoms except normal hydrogen atoms, which have a single proton for a nucleus. Neutrons and protons are about the same size, and both are about 1,800 times larger than the negatively charged electrons that orbit the nucleus.

In a sense, the nuclear arms race began in 1933 in Szilard's mind as he stood at a London stoplight the day after reading Rutherford's warning. Szilard suddenly realized that neutrons might provide the physical key for releasing atomic energy:

If we could find an element which is split by neutrons and which would emit *two* neutrons when it absorbed *one* neutron [original emphasis], such an element, if assembled in sufficiently large mass, could sustain a nuclear chain reaction. . . . In certain circumstances it might become possible to set up a nuclear chain reaction, liberate energy on an industrial scale, and construct atomic bombs. The thought that this might be in fact possible became a sort of obsession with me.[14]

Szilard's obsession grew into a British patent application in the spring of 1934, which included the concept of a nuclear *chain reaction* using neutrons. In the application, Szilard mentioned that uranium and thorium might sustain chain reactions, a guess which proved to be true. He also discussed the concept of *critical mass*. Assuming that the proper element had been found, Szilard showed theoretically that there was a minimum amount of that element, the critical mass, which would be necessary to sustain a chain reaction.[15]

Szilard was alarmed by his own ideas: "Knowing what this would mean—and I knew it because I had read H. G. Wells—I did not want this patent to become public."[16] His biggest fear was that if he were right, the Nazis might use his idea to build atomic bombs and conquer the world.

In order to keep the patent secret, Szilard arranged to assign it to the British Admiralty Office, after the British War Department had turned down his request, saying that "there appears to be no reason to keep the specification secret so far as the War Department is concerned."[17]

In a letter to the Admiralty, Szilard explained why he was following this unusual course:

The object of this Patent has nothing to do with instruments of war, but it contains information which could be used in the construction of explosive bodies based on processes described in the Specification. Such explosive bodies would be very many thousand times more powerful than ordinary bombs, and in view of the disasters which could be caused by their use on the part of certain Powers which might attack this country, it appears very undesirable that such information should be published through the medium of this Patent.[18]

Szilard's action slowed the spread of information that might have led Nazi scientists toward the chain-reaction concept and atomic bombs. But Szilard also knew, as Wells had put it, that "other men would be doing this" within a few years. His only hope, as he wrote the Admiralty, was that Britain would get there first: "information will *leak* out sooner or later. It is in the very nature of this invention that it cannot be kept secret for a very long time, and my only concern is that the processes should be developed in this country a few years ahead of certain other countries."[19]

Szilard did not oppose further work on the chain-reaction concept. If non-Nazi scientists ceased such work, then the Nazis might discover the atomic bomb first. But he explained his concerns privately to other scientists, suggesting that they stop publishing work bearing on the possibility of chain reactions in the *open literature*.[20] In 1935, he suggested setting up a private system of circulating research papers on "the *dangerous zone*." Such work would be kept out of the public journals, and only a select list of scientists in the United States, Great Britain, and "one or two other countries" would receive the private manuscripts.[21] Szilard's suggestion was a deep affront to the scientific tradition of free and open publication, however, and he found almost no support for censoring work that might lead to a chain reaction.

Szilard renewed his campaign in late 1938, spurred on by the dramatic announcement by Otto Hahn and Fritz Strassman that bombarding uranium with neutrons had resulted in the splitting, or fissioning, of the uranium atom into lighter elements. This process released large quantities of energy. Szilard guessed immediately that the fissioning of uranium would release additional neutrons, perhaps enough to sustain a chain reaction. He again was reminded of *The World Set Free:* "All the things which H. G. Wells predicted appeared suddenly real to me."[22]

By this time Szilard had emigrated to the United States. He secured the reluctant cooperation of physicists in the United States and Great Britain to censor their work. They agreed to submit papers to scientific journals accompanied by letters asking that the papers be withheld from publication until further notice. This arrangement soon collapsed when Frédéric Joliot in France refused to cooperate, claiming that crucial information had already appeared in the American press.[23] Even after Joliot's refusal, Szilard argued for self-censorship, until finally he was warned that further agitation might cost him his research privileges at Columbia University.[24] During 1939, there were about 100 papers published on fission.[25]

It was not until the spring of 1940 that Szilard finally forced the U.S. government to set up a secrecy system for reviewing papers on uranium research. He sent a paper to *Physical Review* on the construction of a chain-reacting system composed of graphite and uranium, the same type of system that was used two years later to achieve the world's first chain reaction. (The paper began, "As early as 1913 H. G. Wells forecast the discovery of induced radioactivity for the year 1933 and described the subsequent advent of nuclear transmutations on an industrial scale. . . .")[26] Szilard asked the editor to withhold publication, and then threatened to publish the paper unless the government acted. A letter written at Szilard's request by Albert Einstein discreetly conveyed this threat directly to President Roosevelt.[27] As a result, official censorship was established, and a system of secrecy was instituted that has become a principal characteristic of nuclear development, overriding the system of free and open publication in science which prevailed before the war began. (Szilard's February 1940 article was not even widely circulated during the Manhattan Project to other scientists, due to security rules. The authorities first refused and then delayed its declassification until December, 1946, and the paper was not finally published until 1972.)[28]

While Szilard struggled for voluntary censorship, news of the discovery of fission found its way into print. One of the early accounts appeared in the *New York Times* on February 25, 1939, written by William L. Laurence. Laurence was hired by the Manhattan Project in 1945, and was the only journalist to witness the first atomic explosion in New Mexico in July, 1945. He was also the only reporter to witness the bombing of Nagasaki (from an observer plane), and became one of the world's best-known interpreters of atomic energy after World War II.

On February 24, 1939, Laurence attended an informal session of the annual convention of the American Physical Society at which Neils Bohr and Enrico Fermi described the splitting of the uranium atom. In a book written after the war, Laurence described his reaction to the idea of a chain reaction: "I remember saying to myself, 'This is the *Second Coming of Prometheus*, unbound at last after some half a million years, bringing down a fire from the original flame that had lighted the stars from the beginning.' "[29]

Laurence then wondered whether the Nazis might use the energy from a chain reaction to create powerful new weapons. After the meeting he asked Bohr and Fermi about the likelihood that fission could be used to make weapons. Fermi insisted that such a development was twenty-five to fifty years off, if it proved possible at all.[30]

In his news story the following day, Laurence hailed the fantastic promise of the new discoveries in language that echoed the language of the radium craze. He reported that Bohr and Fermi had announced the

> . . . creation of a half dozen of the heavier elements out of uranium, accompanied by the release of tremendous quantities of energy. . . .
>
> The work on the newest *"fountain of atomic energy"* is going feverishly in many laboratories both here and in Europe. . . . It constitutes the biggest *"big game hunt"* in modern physics . . . marking the most important step yet made by science toward the transmutation of the elements and the utilization of the *vast stores of energy* locked up within the nuclei of atoms.
>
> The new method for the release of atomic energy and the transmutation of the elements is regarded as the nearest approach yet to be made to the finding of a modern version of the *"Philosopher's Stone"* of the alchemists.[31]

A few months later, however, the destructive potential of this alchemy was noted in a *Times* story on the spring meeting of the American Physical Society. Bohr declared that bombarding a small amount of the isotope U-235 with neutrons of the right energy levels could start a "chain reaction" or atomic explosion sufficient to blow up a laboratory and the surrounding country for many miles.[32] (In nature, the element uranium occurs in three different isotopes. Isotopes are species of an element that have different numbers of neutrons in the atomic nucleus. Isotopes are chemically identical, but have different physical properties and atomic weights. About 99.3% of naturally occurring uranium is U-238, 0.7% is U-235, and 0.004% is U-234.)

The *Times* article went on to explain that the major obstacle to setting up a chain reaction was finding a way to separate the U-235 from the more common U-238. Dr. L. Onsager of Yale claimed that such separation was possible. Other physicists responded that the cost would be prohibitive. The article concluded that "if Dr. Onsager's process of separation should work, the creation of a nuclear explosion which would wreck an area as large as New York City would be comparatively easy."[33]

For the most part, however, media accounts were full of mythic imagery of the coming Golden Age. In 1940 *Collier's* published a stunning portrait of the uranium-powered world we would be living in once the war was over. The author, R. M. Langer, was a physicist at the California Institute of Technology, which gave his predictions an extra patina of credibility. Langer predicted that we would soon be living in quiet underground homes using "energy so *cheap* it isn't worth making a charge for it."[34] Agriculture would also be carried on underground, and the surface of the earth given over to parks and wilderness. We would roam the largely roadless landscape in uranium-powered autos equipped with living rooms and giant soft tires to spare the countryside. Everyone would own atom-powered airplanes that could hover like helicopters, soar fifty miles above the earth, and streak from place to place at thousands of miles per hour (see figure 2).[35]

This vision was not, Langer cautioned,

> a promise of *Utopia* centuries away. It is a statement of facts that will profoundly change for the better the daily lives of you and yours.[36]

There would be:

> unparalleled richness and opportunities for all. Privilege and class distinctions and the other sources of social uneasiness and bitterness will become relics because things that make up *the good life* will be so abundant and inexpensive. War itself will become obsolete because of the disappearance of those economic stresses that immemorially have caused it. Industrious, powerful nations and clever, aggressive races can win at peace far more than could ever be won at war.[37]

And lest the reader demur that this vision might be excessive, Langer insisted:

> This is not visionary. The foundations of the *happy era* have already been laid. The driving force is within our grasp. Reality is about to be handed from the scientists in their laboratories to the engineers in their factories for application to your daily life. It is a new form of power—atomic power.[38]

FIGURE 2 *Popular Mechanics* ran this illustration of what life would be like in the "uranium age" in a January, 1941 article. Almost all activities except transportation and recreation would be moved underground. People would roam the parklike surface of the planet in giant rubber-tired vehicles, while they would travel through the air in "flying wings" that are so swift that "no two points are more than seven flying hours apart."

Source: *Popular Mechanics*, January 1941, p. 3. Used by permission.

There were two problems in this atomic utopia: boredom and atomic weapons. Everyone would have "access to everything produced anywhere in the world. . . . But it is harder than ever to provide something interesting, new, unexpected, colorful. . . . They have everything and still they must keep on their toes."[39]

Langer knew that uranium could be destructive. "In the hands of *eccentrics and criminals* who might be capable of using U-235 for destructive purposes, a weapon of extraordinary power will be available."[40] As a result, citizens of the future would have to be "educated better than they are now with respect to their social responsibilities. Certain kinds of *abnormalities*, ordinarily considered lightly, will have to be overcome or their possessors destroyed."[41] And while Langer predicted the production of uranium "at a rate of ten thousand tons a year," he also noted that because of the weapons problem, "society will have to keep track of all the Uranium produced and refined and take action against any individual who tries to accumulate a *dangerous supply*."[42] (This "keeping track" process has since become known as *nuclear safeguards*; see part VI.)

The articles on atomic power published before wartime censorship began sometimes contained information that would be helpful to anyone planning an atomic weapons program. In June, 1940, *Harper's* published an article by John J. O'Neill ("Enter Atomic Power") that made it clear that one route to an atomic bomb involved the difficult engineering task of separating a rare isotope of uranium, U-235, from the more plentiful U-238. Since isotopes are chemically identical, separating the U-235 from the U-238 would involve taking advantage of the slight difference in weight between the two isotopes.

O'Neill described the first experimental separation of the isotopes as the "'*Open Sesame*' to Nature's treasure chest of atomic energy."[43] O'Neill admitted that U-235 could be used to make a very powerful weapon, but felt that it was "*unthinkable* that we should fail to use constructively this new discovery of a way to release atomic energy."[44]

One of the last articles to appear before censorship began was written by William L. Laurence and appeared in the September 7, 1940 issue of the *Saturday Evening Post*. Laurence hailed the "new substance, a veritable *Prometheus* bringing to man a *new form of Olympic fire*,"[45] while casually noting that one pound of the new substance would have the explosive power of 15,000 tons of TNT.[46] He compared the atomic researchers to "many an explorer before them, among whom Columbus is the best example,"[47] coming unexpectedly upon "a miraculous *new continent* of matter, as rich and wonderful in its way as the Americas proved to be years after their discovery,"[48] "the *Promised Land* of Atomic Energy."[49]

Later in the article the researchers were suddenly transformed into "*Little David* . . . cracking nature's *Goliath* in two and forcing him to give up an enormous amount of his strength."[50] This strength, Laurence believed, was such that "a breath of air would operate a powerful airplane continuously for a year, a handful of snow would heat a large apartment house for a year, . . . and a cup of water would supply the power of a great generating station of 100,000 kilowatt capacity for a year."[51]

Laurence's article contained large amounts of specific information about the still dimly understood atomic processes. He accurately described the use of neutrons to penetrate the nucleus and the difficult problem of isotope separation of U-235. Laurence also expressed the fear that the Nazis, with their "totalitarian economy,"[52] would have an easier time mounting the massive technical effort needed to separate isotopes. Curiously, he too concluded that despite the awesome explosive potential of the new substance, it "would not likely be wasted on explosives."[53]

CHAPTER THREE

The Battle of
the Laboratories

It took 495 words to start the Manhattan Project. That was the length of a letter that Leo Szilard persuaded Albert Einstein to send to President Roosevelt in August, 1939. The letter recommended that the United States explore the possibility of producing atomic bombs, thus initiating a process that ultimately led to the secret Manhattan Project, a massive operation with thirty-seven installations in nineteen states and Canada, more than 43,000 employees, and a wartime budget of 2.2 billion dollars.[1] The result of this huge industrial enterprise was what some Manhattan Project participants felt was nothing short of a miracle. As Joseph Hirschfelder, a group leader at the bomb-building laboratory at Los Alamos, said in a 1975 talk:

> I believe in *scientific-technological miracles,* since I saw one performed at Los Alamos during World War II. The very best scientists and engineers were enlisted in the Manhattan Project. They were given overriding priorities. They got everything which they deemed essential to their program; the cost was unimportant. . . . In a period of two-and-a-half years, they produced the miracle—an atomic bomb which creates temperatures of 50,000,000 C. (or 15,000 times as hot as molten iron), and pressures of the order of 20,000,000 atmospheres (or pressure greater than at the center of the earth), while unleashing the tremendous energy stored in the atomic nuclei.[2]

Szilard's concern that the Nazis might be developing atomic weapons led him to contact Einstein; Szilard then arranged for Dr. Alexander Sachs, a financier and friend of President Roosevelt, to present the letter directly to the president. Szilard wrote two drafts of the proposed letter, one short and one long. As he said later, "We did not know just how many words one could put in a letter which a president is supposed to read. How many pages does the fission of uranium rate?"[3]

Einstein signed the longer of the two drafts, which told Roosevelt:

> Some recent work by E. Fermi and L. Szilard, which has been communicated to me by manuscript, leads me to expect that the element uranium may be turned into a new and important source of energy in the immediate

future. Certain aspects of the situation which has arisen seem to call for watchfulness and, if necessary, quick action on the part of the Administration.[4]

The letter first mentioned the possibility of setting up a chain reaction in uranium which would release "vast amounts of power and large quantities of new radium-like elements," a possibility which "appears almost certain" to occur "in the immediate future."[5]

Szilard and Einstein were somewhat more conservative on the outlook for atomic bombs:

> This new phenomenon would also lead to the construction of bombs, and it is conceivable—though much less certain—that extremely powerful bombs of a new type may thus be constructed. A single bomb of this new type, carried by boat and exploded in a port, might very well destroy the whole port together with some of the surrounding territory. However, such bombs might very well prove to be too heavy for transportation by air.[6]

After a recommendation for securing a supply of uranium ore and providing funding for further experiments at university laboratories, the letter concluded with a warning that the Nazis might also be working on a bomb.

Sachs delivered the letter to Roosevelt in October. After studying it for a moment, Roosevelt's reaction was, "Alex, what you are after is to see that the Nazis don't blow us up." Turning to a trusted aide, "Pa" Watson, Roosevelt concluded, "This requires action."[7]

In response to the Szilard/Einstein letter, Roosevelt created the Advisory Committee on Uranium, which in turn became the S-1 Committee. Finally, the Manhattan Project was formed, under the stewardship of General Leslie R. Groves of the Army Corps of Engineers.

A combination of fear of the Nazis and technological problems accounted for the gargantuan size of the Manhattan Project. It was well known that the Nazis had excellent scientists and engineers, ample industrial capacity, and a source of uranium in the Czechoslovakian mines they had captured. A uranium research program was under way at one of the Kaiser Wilhelm institutes under the direction of the son of the German undersecretary of state.[8]

By 1942, scientific evidence indicated that atomic bombs were possible, using either U-235 or plutonium as the fissionable material. Scientists had identified a number of possible methods for producing these materials. But each method presented its own set of unprecedented engineering difficulties, and there was no way to know how fast any of the proposed methods could be made to produce sufficient quantities of U-235 or plutonium.

No large industry had ever tried to expand so rapidly as the Manhattan Project, going from the experimental stage to full-scale industrial production in less than four years. As Groves explained, under normal conditions with such untried technology, there would be a multi-year, three-stage development process: "Its usual sequence is, first, laboratory research, followed by the design, construction, and operation of a semi-works. Only after the semi-works is in successful operation is the design of the commercial plant begun."[9]

This leisurely path was impossible if a weapon was to be produced before the war ended. On the basis of the skimpy laboratory results, engineers drew up plans for full-scale plants, without knowing whether critical engineering and design problems could be solved. Groves wrote that the only way to proceed under such conditions was to assume "that our work would be successful. This was normal procedure. Always we assumed success long before there was any real basis for the assumption; in no other way could we telescope the time required for the over-all project. We could never afford the luxury of awaiting proof of one step before proceeding with the next."[10]

Groves applied this philosophy to three major production technologies for fissionable materials: the separation of U-235 by gaseous diffusion, electromagnetic separation of U-235, and the production of plutonium from uranium irradiated in high-power production reactors. According to the Atomic Energy Commission's official history of the years 1939 to 1946, *The New World:*

> [T]he severe specifications imposed by the need for great reliability in operation, extreme operating conditions, and the magnitude and complexity of the equipment were common to all three approaches. Overcoming corrosion, fabricating metals, purifying materials, understanding thermodynamics, building auxiliary materials, developing special tools, using new materials, and testing models made up the engineers' day throughout the Manhattan Project.[11]

The Manhattan Project constructed two new cities to house the tens of thousands of workers building the production plants. In Tennessee, there was Oak Ridge (also known as *Dogpatch* or *the Patch*), which grew to be the fifth largest city in the state in less than two years, with a population of 79,000. In Washington, Hanford became in one year the fourth largest city in the state, with 60,000 residents.[12]

In addition to these large production plant towns, the Corps also built Los Alamos, a special town to design and manufacture the bomb, located on an isolated mesa in New Mexico outside of Santa Fe. By the end of the war, 5,800 people lived on *the Hill*, as Los Alamos was known by its inhabitants.[13]

Tremendous institutional momentum developed to complete the weapon before the war ended. As the budget for the project grew from its initial authorization of $6,000 in the fall of 1939 to its wartime total of $2.2 billion, the stakes shifted. Groves and his colleagues were no longer simply racing against the Nazis; they were racing against the war itself. If they failed to build a successful bomb in time for use in the war, there was bound to be a grueling congressional investigation of the possible misuse of precious wartime funds. As Groves later wrote:

> I knew ... that if we were not successful, there would be an investigation that would be as explosive as the anticipated atomic bomb. Once, in 1944 [General] Somervell told me with a perfectly straight face, at least for a moment: "I am thinking of buying a house about a block from the Capitol. The one next door is for sale and you had better buy it. It will be convenient because you and I are going to live out our lives before Congressional committees."[14]

At Los Alamos, the fear of failure was expressed in a modified version of Ralph Waldo Emerson's "Concord Hymn":

> From this crude lab that spawned a dud
> Their necks to Truman's axe uncurled
> Lo, the embattled savants stood
> And fired the flop heard around the world.[15]

With the detonation of the atomic bomb, the sprawling scientific and industrial complex justified its existence. The Manhattan Project became the archetypal example of the successful research and development (R&D) crash program. Secretary of War Henry Stimson said the success of the project "probably represents the *greatest achievement* of the combined efforts of science, industry, labor, and the military in all history."[16]

THE HEART OF SECURITY

The Manhattan Project was secret. Its cities were built in secret, its research was done in secret, its scientists traveled under assumed names, its funds were concealed from Congress, and its existence was systematically kept out of the media. The emphasis on secrecy continued after the war ended, and has become a principal characteristic of the nuclear mindset.

General Groves always insisted that secrecy was a means, not an end, in the Manhattan Project: "Security was not the primary object of the Manhattan Project. Our mission was to develop an atomic bomb of such power that it would bring the war to an end at the earliest possible date. Security was an essential element, but not all-controlling."[17]

There were two principal components to Groves's security precautions: the *compartmentalization of information* inside the project, and *press censorship* outside. As Groves outlined it in his book, *Now It Can Be Told,* the problem inside the project was "to establish controls over the various members of the project that would minimize the likelihood of vital secrets falling into enemy hands. Dr. Bush had already expressed concern over the risks incurred through the free exchange of information among the various people on the project. This flow had to be stopped, if we were to beat our opponents in the race for the first atomic bomb."[18]

Compartmentalization, or the restriction of knowledge about various aspects of the Manhattan Project to the "compartments" in which the knowledge was being developed, was central to this strategy. As Groves said, "Compartmentalization of knowledge, to me, was the very *heart of security*. My rule was simple and not capable of misinterpretation—each man should know everything he needed to know to do his job and nothing else."[19]

Groves claimed that this *need-to-know* rule made "quite clear to all concerned that the project existed to produce a specific end product—not to enable individuals to satisfy their curiosity and to increase their scientific knowledge."[20] Some scientists blamed compartmentalization for slowing down the progress of the Manhattan Project, saying it impeded the flow of scientific information. At a Senate hearing in 1945 after the end of the war, Leo Szilard testified that "compartmentalization of information, which was practiced in the atomic-energy project from November 1940 on was the cause of our failure to recognize that light uranium [U-235] might be produced in quantities sufficient to make atomic bombs. . . . We did not put two and two together because the two twos were in different compartments; they were not together."[21]

The most important effect of the security policies was on the discussion of the larger implications of the development of atomic weapons. For example, when J. Robert Oppenheimer, the physicist Groves selected to run Los Alamos, proposed weekly colloquiums at Los Alamos to share information among the different divisions, Groves pressured Oppenheimer to limit the number of participants and to discuss only scientific issues.[22] Historian Martin J. Sherwin wrote in his book *A World Destroyed:* "In an important way, then, the Manhattan Project's security system served—as all such systems inevitably serve—as an extension and reflection of the policymaking process itself. Established to keep the secret of American efforts to develop atomic weapons from the Germans, it was soon transformed into a means for controlling the activities of scientists—and the flow of information about the Project to America's allies."[23]

Under the need-to-know rule, the tens of thousands of workers who built and operated the plants at Oak Ridge and Hanford had no idea what they were working on. Security investigations were run on all employees, and repeated for those in sensitive jobs. Oppenheimer was under heavy surveillance: his phone calls were recorded, and the places he stayed were probably bugged.[24] The Manhattan Project had its own secret police, who came to be known as *creeps*.[25] G-2 (Army Intelligence) and the FBI also participated in security operations.

There were undercover intelligence agents at all of the major installations of the Manhattan Project. Some agents took jobs in the plants to monitor employees. Workers were threatened with large fines or jail sentences for violating security regulations. At one of the uranium separation plants, every employee wore a badge with a number from one to five on it showing his or her security level. Employees at this plant also wore color-coded uniforms based on their security level.[26] Workers at the plant did not know what their dials and meters were registering; they only knew to call for a supervisor when their instruments exceeded specified limits.[27] Scientists who actually knew what was going on were not supposed to discuss their work with their spouses, and many did not. A. H. Compton, who headed the Metallurgical Laboratory of the Manhattan Project, a research facility in Chicago, refused to keep secrets from his wife, and managed to get a security clearance for her as well, although this case was an exception to the rule.[28]

Groves fought the establishment of unions at Manhattan Project facilities because he believed that the existence of unions would compromise security:

> We simply could not allow anyone over whom we did not have *complete control* to gain the over-all, detailed knowledge that a union representative would necessarily gain. Neither could we permit the discussions between workers that would be bound to occur in union meetings. And obviously we could not have security officers present to monitor their meetings. Also, some information would inevitably filter back to the International Brotherhoods of the various unions. . . .[29]

Press censorship complemented compartmentalization. As Groves put it, "Press security was the other side of the coin. Here we had the invaluable cooperation and assistance of the Office of Censorship [of the War Department]."[30] To Groves

> the general principles governing our control of information were simple. First, nothing should be published that would in any way disclose vital information. Second, nothing should be published that might attract attention to any phase of the project. Third, it was particularly important to keep such matters out of any magazine or newspaper that was likely to be read by an enemy agent or by anyone whose knowledge of scientific progress would enable him to guess what was going on.[31]

Although "the press was always a bit restive,"[32] Groves managed to get his way, keeping knowledge of the secret project from the American people until the announcement of the bombing of Hiroshima on August 6, 1945.

Groves was concerned about the security implications of every word: "The Manhattan Project always carefully avoided drawing undue attention to its work and to its people. Code names for our projects were deliberately innocuous."[33]

This concern for words began with the name of the project itself. The original name was the *"Laboratory for the Development of Substitute Ma-*

terials," abbreviated *DSM*. Groves was not in charge when this name was selected, and he tried to have it changed "on the grounds of security, feeling that the name was bound to arouse curiosity."[34] His original complaint was ignored, but on the day that a draft of a general order was to be published announcing the formation of the new engineering district under the designation DSM, Groves again objected, this time successfully. The officer in charge of the project at the time was located in Manhattan, so the project was christened the *Manhattan Engineering District*, although it is now almost universally referred to as the Manhattan Project.[35]

Groves also employed linguistic "camouflage" whenever he could:

> It was in order to prevent any speculative articles as well as the publicizing of any of our efforts that the press and radio had been asked to avoid the use of certain words, such as "atomic energy." Certain *decoy words*, such as "yttrium," were included in the list to camouflage [the] real purpose [of the project].[36]

Groves also kept track of articles published before the war that might be helpful to enemy agents. In 1943, for example, the Manhattan Project contacted the *Saturday Evening Post* about the article William L. Laurence had written in September, 1940. The *Post* was asked to report any requests for that back issue, and to delay sending the issue out until they received further instructions. No requests were received.[37] Groves hired Laurence in 1945 to work for the Manhattan Project. Laurence was shocked when security officials seized the copy of his 1940 *Saturday Evening Post* article that he was carrying in his briefcase, stamped it Secret, and locked it away in a safe.[38]

Groves tried to avoid publicity about himself, so he was quite upset when he heard the translation of the Greek name that G-2, Army Intelligence, had selected for a program to learn about German atomic research. "Imagine my horror, then, when I learned that G-2 had given the scientific intelligence mission to Italy the name of 'Alsos,' which one of my more scholarly colleagues promptly informed me was the Greek word for 'groves.' My first inclination was to have the mission renamed, but I decided that to change it now would only draw attention to it." (In a reassuring footnote in *Now It Can Be Told*, Groves notes that *alsos* actually means *grove*.)[39]

Code names, some of them ludicrously transparent, were assigned to many top scientists. Enrico Fermi became Henry Farmer; Eugene Wigner was Eugene Wagner; A. H. Compton became A. H. Comas,[40] and Ernest Lawrence was renamed Ernest Lawson. According to an article by Daniel Lang written soon after the atomic bombs were dropped on Japan in 1945, Lawrence had to be given a second code name after his first one became too well known. The security people "dreamed up a brand new one—*Oscar Wilde*. It was selected, I was told, because Wilde had written a play called *The Importance of Being Earnest*."[41]

Lang also recorded some of the other common code words used in the project:

> "*top*" for "atom," "*topic*" for "atomic," "*boat*" for "bomb," "*topic boat*" for "atomic bomb," "*urchin fashion*" for "uranium fission," "*spinning*" for "smashing," and "*igloo of urchin*" for "isotope of uranium." The word for U-235 was "*tenure,*" and I learned how perfectly obvious it was from an eager lieutenant. "You see," he told me, "two and three are five and five make ten. Right? O.K., that's where the 'ten' comes in. The 'ure' stands for uranium. So, 'ten' plus 'ure' equals 'tenure.' A cinch."[42]

Plutonium was called *49*, its atomic number, 94, backwards.[43] *Tube Alloys* was the code name for the British version of the Manhattan Project.[44] *Tuballoy* was also used as a code word for uranium. (An official receipt for the Hiroshima bomb records the arrival on Tinian Island in the Pacific of a "*projectile unit* containing [blank] kilograms of *enriched tuballoy* at an average concentration of [blank]." Also on the receipt is a later note from the pilot of the Enola Gay, the plane which dropped the Hiroshima bomb, saying that "the above material was *expended* to the city of Hiroshima, Japan at 0915 6 Aug.")[45] The atomic bomb was referred to as a *gadget* or a *device*; the latter term is still used today, as in the phrase, *peaceful nuclear device*.

Nowhere in the Manhattan Project was security tighter than at Los Alamos, where the actual design and construction of the world's first atomic bombs took place. Robert Oppenheimer wanted a site so secure that he could relax Groves's rule about compartmentalization and allow for more open discussion among the scientists working on the bomb.[46]

Los Alamos was not on the map; unlike Hanford and Oak Ridge, which could be found in post-office directories, Los Alamos had no post office. Mail for people working there was addressed to Post Office Box 1663, Santa Fe, New Mexico.[47] Newspapers near Oak Ridge and Hanford were allowed to print social notes about people working at the two plants, but security officers tried to prevent newspapers from even mentioning the existence of Los Alamos.[48] When Oppenheimer began to recruit scientists for the bomb laboratory, he could not tell them where they were going. He could only say that they would be working on a secret project to build a "*Projectile S-1-T.*"[49]

Phone conversations at Los Alamos were monitored, and censors would interrupt any conversation that appeared to be heading in the wrong direction. Mail was censored and returned for correction or deletion if necessary. There was a rule book for letter writers which proscribed certain words. According to a 1975 talk by Bernice Brode, who worked at Los Alamos from 1943 to 1945, "We could not mention last names, give distances or places nearby, and the *worst word*, 'physicist,' was strictly forbidden. I might say we could write 'theoretical' or 'experimental,' and the censors wouldn't know, but our friends would."[50] Los Alamos was the only place in the United States where mail was censored. (Letters with the censor's stamp on them are now collector's items.)[51]

People were allowed a monthly visit to Santa Fe or one of the other nearby towns, where their movements were closely monitored by plainclothes intelligence agents. According to one woman who lived at Los Alamos, Santa Fe was

> full of men from G2; you could always spot them because they wore snap brimmed hats—straw in the summer—felt in the winter. G2 followed Hill people around town to see that they didn't speak to anyone on the street. People from Los Alamos were supposed to cut their own parents if they met them on the street. G2 saw to it they didn't mail any letter surreptitiously and tailed them. . . .[52]

When press censorship finally ended, the *Santa Fe New Mexican* heaved "a sigh—sigh nothing; it's more of a groan—of relief"—in a front page story entitled "Now They Can Be Told Aloud, Those Stoories [sic] of 'The Hill.'" Reporters had heard dozens of rumors about what Los Alamos was up to, from gas warfare to jet propulsion to rockets to atomic bombs to "a Republican internment camp."[53]

The *New Mexican*'s reporter described the frustrations of a newspaper forbidden to cover the mysterious goings-on in its own backyard: "A whole social world existed in nowhere in which people were married and babies were born nowhere. People died in a vacuum, autos and trucks crashed in a vacuum. . . ."[54]

The paper could not even acknowledge the existence of the numerous explosions that were heard coming from the direction of Los Alamos. (The explosions were part of the test of the chemical explosives that would be used to initiate the much more powerful atomic explosions.) When the chemical explosions occurred, "this paper's phone would ring but the whole staff could just 'no speak English.'"[55]

A VERY BIG FISH

The work of the laboratory at Los Alamos culminated with the detonation of the world's first atomic bomb, at 5:30 A.M. Mountain War Time, July 16, 1945, at Alamogordo, New Mexico, a desert site about 200 miles from Los Alamos. The explosion released energy equivalent to that released by 20,000 tons of TNT;[56] the blast was seen for 250 miles; and the sound was heard for 50 miles (see plate A).[57] For some observers, the explosion also released the metaphors of destruction that the rest of the world would not learn about for three more weeks, with the bombing of Hiroshima.

The code name for this first explosion was *Trinity*, chosen by Oppenheimer. The name grew out of Oppenheimer's recent reading of one of John Donne's *Holy Sonnets*. In the sonnet, which seethes with violent imagery, Donne asks a "three-personed God" to batter his heart, to use its terrific force to "break, blow, burn, and make me new":

> Batter my heart, three-personed God; for You
> As yet but knock, breathe, shine and seek to mend;
> That I may rise and stand, o'erthrow me, 'and bend
> Your force to break, blow, burn, and make me new.
> I, like an usurped town, to'another due,
> Labor to'admit You, but O, to no end;
> Reason, Your viceroy'in me, me should defend,
> But is captived, and proves weak or untrue.
> Yet dearly' I love You, 'and would be lovéd fain,
> But am betrothed unto Your enemy.
> Divorce me, 'untie or break that knot again;
> Take me to You, imprison me, for I,
> Except You'enthrall me, never shall be free,
> Nor ever chaste, except You ravish me.

At the test site, the bomb was placed at the top of a 100-foot steel tower.[58] The location was called *Point Zero* or *Zero*. The observers waited in bunkers several thousand yards away, equipped with dark glasses to view the early stages of the blast. According to Edward Teller, the principal inventor of the hydrogen bomb, "We were told to lie down on the sand, turn our faces away from the blast, and bury our heads in our arms. No one complied. We were determined to look the beast in the eye." Teller even came equipped with suntan lotion.[59]

One of the most graphic descriptions of the effects of the Trinity test is General James Farrell's report, contained in General Groves's official report on the event to the War Department. According to Farrell:

> We were reaching into the unknown and we did not know what might come of it. It can safely be said that most of those present were praying and praying harder than they had ever prayed before. . . .
> . . . when the announcer shouted "Now!" . . . there came this tremendous burst of light followed shortly thereafter by the deep growling roar of the explosion. . . .
> The effects could well be called unprecedented, magnificent, beautiful, stupendous and terrifying. No man-made phenomenon of such tremendous power had ever occurred before. The lighting effects beggared description. The whole country was lighted by a searching light with the intensity many times that of the midday sun. It was golden, purple, violet, gray and blue. It lighted every peak, crevasse, and ridge of the nearby mountain range with a clarity and beauty that cannot be described but must be seen to be imagined. It was that beauty the great poets dream about but describe most poorly and inadequately. Thirty seconds after the explosion, came, first, the air blast, pressing hard against the people and things, to be followed almost immediately by the strong, sustained, awesome roar which warned us of doomsday and made us feel that we puny things were blasphemous to

0.006 SEC.
N   100 METERS

0.025 SEC.
N   100 METERS

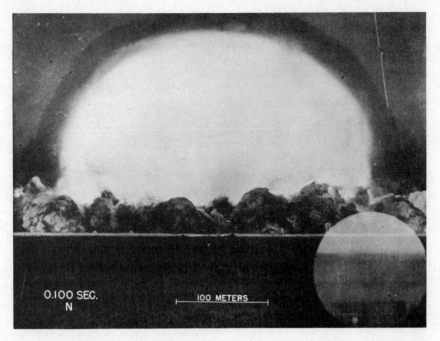

PLATE A   Sequence of photos from Trinity, the first atomic bomb test, July 16, 1945.
Source: Los Alamos Scientific Laboratory.

dare tamper with the forces heretofore reserved to the Almighty. Words are
inadequate tools for the job of acquainting those not present with the
physical, mental, and psychological effects. It had to be witnessed to be
realized.[60]

Oppenheimer told William L. Laurence that he was reminded of a line
of Hindu poetry from the Bhagavad-Gita, "I am become Death, the Shatterer of Worlds."[61] Kenneth Bainbridge, who was in charge of the test, had a
more prosaic reaction. He told Oppenheimer, "Now we are all sons of
bitches."[62] Bainbridge later wrote that Oppenheimer told one of his
daughters that Bainbridge's remark "was the best thing anyone said after
the test."[63]

Groves had taken special precautions to keep the story of the explosion
out of the press, especially on the East and West coasts. Laurence, who,
although working for the Manhattan Project, was still on the *New York
Times* payroll, prepared four different press releases, differing only in the
size of the explosion they described. Laurence numbered every word in the
releases so that, as Groves explained, "it was a simple matter to alter it
without disclosing any secrets to an unauthorized listener-in."[64] (Laurence
had warned the *Times* in a July 12 letter that "zero hour for the revelation
of the *Big Secret* was approaching." He wrote his old editor to make sure
enough space would be reserved for the story:

The story is much bigger than I could imagine, fantastic, bizarre, fascinating and terrifying. When it breaks it will be an eight-day wonder, a sort of *Second Coming of Christ* yarn. It will be one of the big stories of our generation and it will run for some time. It will need about twenty columns on the day it breaks. This may sound overenthusiastic, but I am willing to wager you right now that when the time comes you will agree that my estimate is on the conservative side.)[65]

The explosion occurred at 5:30 A.M., about 1½ hours behind schedule, so that many people were awake when the blast lit up the sky in the west. By 11 A.M., an officer stationed by Groves at the Albuquerque office of the Associated Press reported that unless the Army had something to say, the AP would run its own story on the mysterious explosion. At that point, Groves made the appropriate corrections in one of the releases, which was then issued by the Commanding Officer of the Alamogordo Army Air Base, who knew nothing about what was going on other than that he had been ordered to issue the press release. The release read:

ALAMOGORDO, N.M. July 16

The commanding officer of the Alamogordo Army Air Base made the following statement today:

"Several inquiries have been received concerning a heavy explosion which occurred on the Alamogordo Air Base reservation this morning.

"A remotely located ammunition magazine containing a considerable amount of high explosives and pyrotechnics exploded.

"There was no loss of life or injury to anyone, and the property damage outside of the explosives magazine itself was negligible.

"Weather conditions affecting the content of gas shells exploded by the blast may make it desirable for the Army to evacuate temporarily a few civilians from their homes."[66]

The reference to gas shells was designed to cover the possible evacuation of civilians if the radioactive fallout from the test endangered people outside the test site.

This release did not keep the story out of the local papers, but the Office of Censorship managed to keep it out of the East Coast papers, except for a few lines in one paper. The explosion was covered by West Coast radio, but of course no one had any way of knowing that an atomic bomb was responsible.

A few special messages went out from Oppenheimer and Groves telling others of the explosion. Oppenheimer cabled one Manhattan Project scientist, "You'll be interested to know that we caught *a very big fish*."[67] Oppenheimer's message to his wife if the test was successful was more homely: "You can change the sheets."[68]

The most important messages went from Groves to President Truman, who was meeting with Churchill and Stalin at Potsdam, Germany. Truman needed to know whether the test was successful as soon as possible in order to use the information in the negotiations. Indeed, the meeting had been scheduled in part around the atomic bomb test schedule.[69] The first message to reach American negotiators was promising:

> Operated on this morning. Diagnosis not yet complete but results seem satisfactory and already exceed expectations. Local press release necessary as interest extends great distance. Dr. Groves pleased. He returns tomorrow. I will keep you posted.[70]

A second cable followed, with more complete information:

> Doctor has just returned most enthusiastic and confident that the little boy is as husky as his big brother. The light in his eyes discernible from here to Highhold and I could have heard his screams from here to my farm.[71]

This message told Truman that the test had exceeded all expectations: Highhold was 250 miles from Washington, and the farm was 50 miles away.[72]

HIROSHIMA

At 11 A.M., August 6, 1945, at the White House, President Harry Truman's press secretary began reading one of the most important press releases in history. It was also one of the most carefully worded, having gone through several drafts and having been reviewed by the highest officials of both the United States and Great Britain.[73]

There was no indication to the press that anything extraordinary was afoot. The release began:

> Sixteen hours ago an American airplane dropped one bomb on Hiroshima, an important Japanese army base. That bomb had more power than 20,000 tons of T.N.T. It had more than two thousand times the blast power of the British 'Grand Slam' which is the largest bomb ever yet used in the history of warfare.[74]

According to General Groves, "As the words 'more than twenty thousand tons of TNT' came out of his [the press secretary's] mouth, there was a tremendous rush of reporters for the releases, which were on a table at the exit from the room, and then to the telephones and their offices."[75] Groves was pleased to see that in the ensuing news coverage, "most newspapers published our releases in their entirety. This is one of the few times since government releases have become so common that this has been done."[76]

Truman's press statement was aimed at the Japanese government and its people: "The force from which the sun draws its power has been loosed against those who brought war to the Far East."[77] Unless the Japanese accepted the U.S. surrender terms, there would follow "*a rain of ruin* from the air, the like of which has never been seen on this earth."[78]

Truman also praised the Manhattan Project, declaring that "what has been done is the *greatest achievement of organized science* in history."[79] Comparing the scientists of the Manhattan Project to war heroes, Truman said, "The *battle of the laboratories* held fateful risks for us as well as the

PLATE B Bomb casings similar to those used in the bombings of Hiroshima and Nagasaki. Top: Little Boy bomb similar to the one that destroyed Hiroshima. Bottom: Fat Man bomb like the one dropped on Nagasaki.

Source: Los Alamos Scientific Laboratory.

battles of the air, land and sea, and we have now won the battle of the laboratories as we have won the other battles"[80] (see plate B).

The statement was not specific about how much damage had been done at Hiroshima, an omission that pained Groves. In the hours before the press conference, he had tried frantically to obtain a damage report from U.S. forces in the Pacific. But the firestorm caused by the explosion had completely obscured the ground from aerial reconnaissance. Groves wrote that he "finally compromised by making a minor change in the first paragraph, which I felt would not lessen the impact of the news on the Japanese, and would still leave us a loophole in case the bomb had not the anticipated destructive force."[81]

CHAPTER FOUR

Atoms for Peace

The hideous power of nuclear weaponry generated intense hope that the power of the atom would also produce great benefits for humanity. In his 1963 book, *Change, Hope and the Bomb*, David Lilienthal, the first chairman of the Atomic Energy Commission, talked about the mindset that led to these hopes:

> The basic cause, I think, was a conviction, and one that I shared fully, and tried to inculcate in others, that somehow or other the discovery that had produced so terrible a weapon simply *had* [original emphasis] to have an important peaceful use. Such a sentiment is far from ignoble. We are a peace-loving people. Everyone, our leaders and laymen and scientists and military men, wanted to establish that there is a beneficial use of this great discovery. We were grimly determined to prove that this discovery was not just a weapon. This led perhaps to wishful thinking, a wishful elevation of the *"sunny side" of the Atom.*[1]

Even as fires still shrouded Hiroshima in smoke, Secretary of War Stimson announced in an August 6, 1945 press statement that atomic fission held

> *great promise for sweeping developments* by which our civilization may be enriched when peace comes. . . . With the evidence presently at hand, . . . it appears inevitable that many useful contributions to the well-being of mankind will ultimately flow from these discoveries when the world situation makes it possible for science and industry to concentrate on these aspects.[2]

The conflicting images of the nuclear future were captured in the cliché of the nuclear *crossroads*. As William L. Laurence, who became an enthusiastic promoter of the sunny side of the atom, wrote in a 1948 article in the *Woman's Home Companion*, "Today we are standing at a major *crossroads*. One fork of the road has a signpost inscribed with the magic word *'Paradise'*; the other fork also has a signpost bearing the word *'Doomsday.'*"[3] Laurence trusted that humanity would find a way to avoid going down the path marked "Doomsday." Using nuclear energy, he argued, humanity had "a chance to enter a new *Eden*, . . . abolishing disease and poverty, anxiety and fear." We might "learn to control weather and heredity," and "find the key to the *riddle of old age*."[4] Everything would get better and better; there would be "better, finer and more nourishing plants, better, cheaper, and more abundant fertilizer; better and richer soils, farms, and gardens; better metals and machines; better and finer clothing and homes; *better men and women*."[5] Nuclear plants would pump underground water to turn the world's deserts into "blooming gardens," turning swamps and jungles into "vast new lands flowing with *milk and honey*." Summing up this "*turn-ingpoint* in the history of civilization," Laurence claimed: "Such power plants could, in short, make the dream of the earth as a *Promised Land* come true in time for many of us already born to see and enjoy it."[6]

One of the favorite clichés for describing the ambiguous promise of nuclear energy was the *genie and the bottle*, taken from the story of the fisherman and the genie in the *Arabian Nights*. Walt Disney Productions seized on this fable to symbolize the story of the atom—a story with "a straightforward plot and a simple moral"[7]—in the 1956 book (and film of the same name) *Our Friend the ATOM*, written by Dr. Heinz Haber.

In the Disney version, the unsuspecting fisherman opens the bottle to reveal a ferocious genie, depicted with a swollen, purple-colored body, glaring yellow eyes, enormously long fingernails, and a scowl pierced by pointed fangs. The genie, angered at having been entrapped for so long, prepares to dispatch the fisherman, but the fisherman does some quick talking and tricks the genie back into the bottle. Dismayed, the genie promises to grant three wishes, and generally to behave, if only the fisherman will release him again. This time the genie appears bowing low toward his new master, his hands extended in a gesture of supplication, with a smile instead of a scowl.

Like the fisherman, Haber wrote, scientists had released a terrible genie who threatened to destroy us "with the most cruel forms of death: death from searing heat, from the forces of a fearful blast, or from subtly dangerous radiations. . . . We must bestir our wits. We have the scientific knowhow to turn the Genie's might into peaceful and useful channels. He must at our beckoning grant three wishes for the good of man."[8]

The solution to controlling *"the atomic Genie"* was said to be the nuclear reactor. As for the three wishes, they were for "Power," "Food and Health," and "Peace." As Haber put it, since atomic energy could destroy civilization, "Our last wish should simply be for the atomic Genie to remain forever our friend!"[9] And while the book opens with a double-page spread of a gigantic atomic explosion destroying a city, it ends with a futuristic city, floating in the clouds.

Bringing the atomic genie under control would require a strict separation of the peaceful uses of atomic energy from its military uses. But how could this be accomplished? In his 1914 novel *The World Set Free*, H. G. Wells predicted that the only way to separate these uses was through rigorous international control: a world government that would own the raw materials used to make atomic bombs. The first act of the world government (following the devastating nuclear war that brought it into being) was to seize every atom of the fictional fissionable element *Carolinum*. With the threat of war permanently removed, humankind could then go forward to enjoy the many benefits Wells expected from the peaceful development of atomic energy.

The first official U.S. government report on nuclear development, released in March, 1946, reached the same conclusion as Wells: the only way to guarantee peaceful nuclear development was to create an international *"Atomic Development Authority."*[10] The first act of that authority should be "to bring under its *complete control* world supplies of uranium and thorium."[11] (Thorium, while not fissionable itself, can be transmuted in nuclear reactors to an isotope of uranium, U-233, that can be used to make nuclear weapons.)

The report was prepared by a board of consultants headed by David Lilienthal, who was then the chairman of the Tennessee Valley Authority. The five-member board also included J. Robert Oppenheimer, the former head of Los Alamos, and high-level executives from New Jersey Bell Telephone, General Electric, and Monsanto Chemical. Their report was prepared for the State Department's Committee on Atomic Energy, headed by Under Secretary of State Dean Acheson, and the document became known as the Acheson-Lilienthal report.

The report concluded that the "development of atomic energy for peaceful purposes and the development of atomic energy for bombs are in much of their course interchangeable and interdependent."[12] National governments could not be trusted with nuclear development, since at any time, a nation pursuing an ostensibly peaceful program might convert its fissionable materials to the making of bombs.

The Acheson-Lilienthal report proposed to classify nuclear activities into two categories, *safe* and *dangerous*, and to require that all *dangerous activities*—those which allowed access to material that could be used in weapons—be carried out exclusively by the international Atomic Development Authority. Dangerous activities included mining and processing uranium and thorium ores, and the operation of plutonium production reactors or isotope separation plants.[13] The report further concluded that so long as dangerous activities were being conducted by individual nation states, no system of international agreements and policelike inspections of nuclear facilities could prevent the military use of atomic technology.[14]

The growing Cold War between the United States and the Soviet Union doomed all attempts at international control of atomic energy in the 1940s. From 1946 to 1953, the United States' nuclear program concentrated almost exclusively on expanding the country's atomic arsenal. This enormous weapons-building program was carried out by the Atomic Energy Commission (AEC), which took over the facilities and operations of the Manhattan Project in 1947.

The AEC had been established by the Atomic Energy Act of 1946, which gave the agency control over the production of fissionable material and over all information concerning atomic energy. To watch over this new agency, the act provided for a new congressional committee, the Joint Committee on Atomic Energy (JCAE).

The AEC moved rapidly to improve the country's nuclear weapons capabilities. The agency's budget went from $318 million in 1947 to $1,766 million in 1952 (see figure 3).[15] The program got a sudden boost after September, 1949, when the AEC detected a Soviet nuclear weapons test.

After the Soviet test, President Truman approved a crash program to develop the *hydrogen bomb,* or *Super*, as it was known at the time. Edward Teller, who had pursued the idea of the Super during the Manhattan Project, believed that an atomic bomb could be used as a trigger to initiate a fusion reaction in very light atoms, such as hydrogen. Intense heat and pressure would fuse these light atoms together into heavier ones, releasing huge amounts of energy. The Super was expected to be about a thousand times more powerful than the Hiroshima atomic bomb. By the fall of 1953, the U.S. and the USSR were both well on the way to producing operational hydrogen bombs.

THE MAGIC WAND

On December 8, 1953, President Eisenhower started the *Atoms for Peace* program with a dramatic speech at the United Nations. Eisenhower announced that the United States wanted to share the benefits of nuclear technology with the rest of the world. Beginning by noting that he was speaking in a *"new language ... the language of atomic warfare,"* Eisenhower confronted the *"dark chamber of horrors"* created by the success of the U.S. and USSR at weapons making. But, Eisenhower declared, "this greatest of destructive forces can be developed into a *great boon* for the *benefit of all mankind."*[16] Eisenhower's message turned the phrase *Atoms for Peace* into a slogan of almost religious significance, capturing the hope that humanity would find a way to transcend the destructive uses of nuclear energy and utilize the power of the atom to realize the dream of unlimited material abundance.

The speech marked a major shift in U.S. atomic energy policy. Eisenhower proposed the creation of an international atomic energy authority that would distribute fissionable material to member nations and would

> devise methods whereby ... fissionable material would be allocated to serve the *peaceful pursuits* of mankind. Experts would be mobilized to

FIGURE 3   AEC BUDGETS, 1947–1952 (MILLIONS OF DOLLARS)

	1947	1948	1949	1950	1951	1952
Procurement and production of nuclear materials	$167.4*	$141.0*	$110.6	$168.5	$188.3	$278.3
Weapons development and fabrication			92.9	112.0	163.6	229.2
Development of nuclear reactors			19.3	31.5	44.5	64.4
Research in chemistry, metallurgy, and physics	24.5*	53.4*	26.1	28.9	29.8	34.7
Research in cancer, biology, and medicine			15.2	17.8	21.3	24.5
Community operations—net	18.9	23.7	25.6	19.9	17.3	16.4
Administrative expenses	16.0	24.8	25.2	22.9	24.5	31.4
Other expenses and income—net	32.4	85.3	61.4	13.3	5.3	5.3
Plant construction	59.1	134.4	255.5	256.1	459.2	1,082.2
Totals	$318.3	$426.6	$631.8	$670.9	$953.8	$1,766.4

*Separate figures for these years not available.

Source: Frank G. Dawson, *Nuclear Power, Development and Management of a Technology* (Seattle: U. of Washington Press, 1976), p. 21.

apply atomic energy to the needs of agriculture, medicine, and other peaceful activities. A special purpose would be to provide *abundant electrical energy* in the power-starved areas of the world. Thus the contributing powers would be dedicating some of their strength to serve the needs rather than the fears of mankind.[17]

The successful functioning of the international authority Eisenhower proposed depended on inspections of nuclear facilities and international agreements—exactly the kind of system that the Acheson-Lilienthal report had said would not be capable of stopping the military use of atomic energy. Eisenhower's proposal resulted in the creation of the International Atomic Energy Agency (IAEA). Under the IAEA, materials and facilities that can be used to manufacture nuclear bombs have spread around the world.

In keeping with the new policy, Congress passed the Atomic Energy Act of 1954, which was designed to encourage the commercial development of atomic energy. The Atomic Energy Commission presided over the nation's nuclear development program until the mid-1970s. As well as manufacturing bombs, the AEC conducted an extensive research and development program, running a system of national laboratories that included the Los Alamos Scientific Laboratory, the Oak Ridge National Laboratory, the Hanford Reservation, the Savannah River Plant in South Carolina, the Lawrence Livermore Laboratory in California, and the Brookhaven National Laboratory on Long Island.

Conflict over the AEC's dual role as both promoter and regulator of nuclear power led to the agency's dissolution in 1974. Its functions—and most of its staff—were split into two separate agencies: the Energy Research and Development Administration (ERDA) was assigned the task of developing nuclear weapons and nuclear power; and the Nuclear Regulatory Commission (NRC) was given authority to regulate commercial nuclear power.

In 1977, the Joint Committee on Atomic Energy, which had failed to regulate the AEC, was dissolved, and its duties were dispersed to a number of other congressional committees. When President Carter reorganized the nation's energy agencies, ERDA was absorbed into the Department of Energy (DOE). Though its name does not suggest this, the *Department of Energy* now manages the nation's nuclear arsenal; approximately one-third of the DOE's budget goes to nuclear weapons research and production. The DOE also inherited the AEC's power to control and classify all information on nuclear energy, and it maintains strict secrecy (see chapter 5).

Eisenhower's *Atoms for Peace* speech gave new life to the hope for a nuclear utopia. There would be nuclear powered planes, trains, ships, and rockets; nuclear energy would genetically alter crops and preserve grains and fish; nuclear reactors would generate cheap electricity. And later, under *Project Plowshare*, even nuclear weapons would be put to peaceful uses, turning nuclear swords into plowshares: they would be used to dig harbors and canals, move mountains, and blast loose valuable mineral deposits.

AEC Chairman Lewis L. Strauss captured the expansiveness of the atoms-for-peace vision in a speech to the National Association of Science Writers in 1954:

> Transmutation of the elements—*unlimited power*, ability to investigate the working of living cells by tracer atoms, the secret of photosynthesis about to be uncovered—these and a host of other results all in 15 short years [from 1954]. It is not too much to expect that our children will enjoy electrical energy *too cheap to meter*—will know of great periodic regional famines only as matters of history—will *travel effortlessly* over the seas and under them and through the air with a minimum of danger and at great speeds—and will experience a lifespan far longer than ours, as disease yields and man comes to understand what causes him to age. This is the forecast for an *age of peace*.[18]

The ground-breaking ceremony for the first commercial power reactor, built in Shippingport, Pennsylvania, started with the wave of a *magic wand*. Work on this plant, which was adapted from a reactor designed for the first nuclear submarine, had begun in 1954, generating large amounts of publicity.

President Eisenhower waved the wand over a counting device in a Denver television studio (see plate C). As *Life* magazine described the scene, "With a wave of a *radioactive wand*, President Eisenhower . . . transformed the *bright hope* for atomic power peaceably used into a *solid certainty*. . . . When the counter's needle swung across the dial, it electrically set in motion, 1,300 miles away at the Shippingport, Pa. plant site, an automatically controlled power shovel which scooped up the first symbolic shovelful of earth."[19]

A second great moment occurred on July 18, 1955, when for the first time a nuclear reactor fed power into a utility grid. (The reactor, known as the *Submarine Intermediate Reactor*, or *SIR-A*, had been designed by General Electric as a prototype for a reactor to power the second atomic submarine.) The closing of the switch sending power into the grid was dramatically staged (see plate D). AEC Chairman Strauss told the crowd:

> Before me stands a large, *two-way switch*. If I throw the blade in one direction, it will turn the propeller shaft of a military weapon. But when I throw it in the other direction, as I am about to do, it will send atomic electric power surging through transmission lines to towns and villages, farms and factories—power not to burst bombs or propel submarines, but to make life easier, healthier, and more abundant. This switch is a symbol of the great dilemma of our times. I throw it now to the side of the *peaceful atom*. . . .[20]

The switch-throwing ceremony was part of a summer-long public-relations build-up for the peaceful atom. In late July, the U.S. Postal Service issued an Atoms for Peace stamp (see figure 4). In August the United Nations sponsored an International Conference on the Peaceful Uses of Atomic

PLATE C Brandishing a radioactive magic wand, President Eisenhower, standing in a Denver TV studio on Labor Day, September 6, 1954, activated an automated power shovel 1,300 miles away in Shippingport, Pennsylvania, to begin construction on the first commercial nuclear reactor in the United States.

PLATE D   AEC Chairman Lewis Strauss throwing the switch, on July 18, 1955, "to the side of the peaceful atom."

Energy, in Geneva. Seventy-two countries were represented at the conference, and thousands of scientists from all over the world attended. In the opening address, U.N. Secretary General Dag Hammarskjöld declared that the conference

> might well mark the beginning of a new phase during which man will have left his bewilderment and his fear behind and will begin to feel the elation of one of the greatest conquests made by his mind. The exploitation of nuclear energy for social and economic ends will be a considerable relief from the *oppressive thought* that, in unlocking the atom, we had done no more than unlock the most sinister *Pandora's box* in nature. This, in itself, will have a great *psychological value* and should free our *best creative efforts....* I am sure that this Conference will demonstrate the many practical uses to which those discoveries could be put for curing some of our worst physical, social, and economic ills, for raising the standard of living, and for lifting mankind to a higher level of well-being.[21]

The AEC hastened to build a number of new atomic toys. The commission had an early interest in *nuclear-powered planes*, and the *Nuclear Energy Propulsion for Aircraft (NEPA)* project was begun in 1948. Atomic-powered airplanes would make long-distance bombing easier, since the planes were expected to be able to circle the globe without refueling. As late as 1959, the Joint Chiefs of Staff were assuring Congress that "there is *considerable military potential* in the nuclear powered aircraft and that early achievement of the capability for nuclear flight would be in the *national interests.*"[22] But in 1961, citing *"chronic overoptimism"* and unavoidable public-relations problems, Secretary of Defense Robert S. McNamara cancelled the program. McNamara told Congress that a nuclear aircraft would "expel some small fraction of radioactive fission products into the atmosphere, creating an important *public relations problem* if not an *actual physical hazard.*"[23] The aircraft reactor program cost over $1 billion (in 1950s dollars), and no reactor-powered plane ever flew.[24]

Nuclear developers were also enthusiastic in the 1950s and 1960s about another flying reactor program: *nuclear-powered rockets*. This plan called for using nuclear reactors to provide heat that would propel a gas through a rocket nozzle, producing more thrust than a chemical rocket of equivalent size. The nuclear rocket program began at Los Alamos in 1955 with the development of the *Kiwi reactors*, inauspiciously named after the flightless New Zealand bird. Experiments with Kiwi reactors provided the technology necessary to develop a *Nuclear Engine for Rocket Vehicle Application*, or *NERVA*. The development plan then called for a *Reactor in Flight Test*, or *RIFT*. There was great confidence that this program would produce a flying rocket reactor. In 1958, JCAE Chairman Clinton Anderson said that there was *"ample scientific evidence* that nuclear propulsion offers the best hope for propulsion of a space vehicle with its powerful and long-lived fuels and tremendous power potential."[25]

FIGURE 4 "ATOMS FOR PEACE" FIRST-DAY COVERS Two contrasting views of President Eisenhower's "Atoms for Peace" campaign, as seen by companies specializing in designing embossed cachets for envelopes for first-day-of-issue stamp collectors. The first depicts a nuclear-powered paradise, while the other shows a mushroom cloud.

According to an editorial in a 1961 special edition of *Nucleonics*, the United States was "on the *threshold of an era* which some are already calling 'The Nuclear Space Age'—the joining in inevitable matrimony of two of contemporary man's most exciting frontiers, nuclear energy and outer space."[26] *Nucleonics* argued that military competition with the Soviet Union made it essential to rush the atom and outer space to the altar. Nuclear rockets were said to offer "the *only realistic possibility* for overcoming the space payload lead of the Soviet Union. . . . In short, the nuclear rocket gives this country its only chance to catch up with—indeed to surpass—the USSR."[27] Neither the Air Force nor the Navy had formally requested nuclear rockets by 1961, a situation which *Nucleonics* deplored. There would surely be an "easier flow of development dollars" if there were "a clear-cut military requirement."[28]

Rocket reactors would create unique radiation hazards, given that a highly radioactive reactor could come crashing down into a populated area. But a 1962 report from Los Alamos claimed that this problem could be handled: "A *vigorous radiological safety program* is being conducted concurrently with engineering development such that *adequate knowledge* of the potential hazards of nuclear propulsion systems will be available to *assure* their *safe use* by the time *flyable reactors* are available."[29]

By the late 1960s, both the Kiwi reactors and the prototype NERVA had been tested at the *Nuclear Rocket Development Station* in Nevada, and the AEC and the National Aeronautics and Space Administration were spending about $100 million a year on the project.[30] Nevertheless, the estimated date of an actual flight test continued to recede indefinitely into the future. As the staff director of the Senate Committee on Aeronautics and Space Sciences pointed out in 1964, there was already "a *long engagement* without setting '*the marriage date.*' . . . The price of *never-ending wooing* may be too high for one to continue the romance."[31] In 1972, the AEC was finally forced to break off the engagement, admitting that the purported advantages of the rocket reactor could not be shown to justify the expenditure of the hundreds of millions of dollars more that would be necessary to produce an actual, flying rocket reactor.[32]

The U.S. government also funded a radically different nuclear-powered rocket program in the late 1950s and early 1960s. This wild scheme, known as *Project Orion*, called for propelling a spacecraft into orbit and out into the solar system with a series of nuclear bomb explosions. Using an apparatus modeled on coin-operated Coke machines to move the bombs from storage to the launch position, the craft would use bursts of compressed air to whisk the bombs to about 100 feet from the ship, where they would detonate. The shock wave from each successive blast would slam into a giant *pusher plate*, sending the ship lurching through the cosmos at a top speed of 100,000 miles per hour.[33] NASA called Orion's bombs "*pulse units,*" while the Air Force referred to them as "*charge propellant systems.*"[34]

According to one participant in Project Orion, physicist Freeman Dyson, the scientists and engineers who began working on the project saw it as a way of turning nuclear weapons to peaceful ends. In 1958, when the project was formally established, Dyson wrote a "Space Traveler's Manifesto" which described these hopes: "We have for the first time imagined a way to use the huge stockpiles of our bombs for better purpose than for murdering people. Our purpose, and our belief, is that the bombs which killed and maimed at Hiroshima and Nagasaki shall one day *open the skies* to man."[35] In his 1979 book *Disturbing the Universe,* Dyson said that the Orion team was not particularly concerned with the fallout which its bomb-powered rocket would leave in its wake:

> It was possible for us in 1958 to enjoy the thought of leaping into the sky with a trail of nuclear fireballs glowing behind us, because at that time the United States and the Soviet Union were testing bombs in the atmosphere at a rate of many megatons per year. We calculated that even our most amibitious program of Orion flights would add only about one percent to the contamination of the environment that the bomb tests were then causing.[36]

(Dyson wrote that his enthusiasm "rapidly cooled" after he learned more about fallout and the biological effects of radiation.)[37]

Dyson and his co-workers flight-tested a scaled-down model of their spaceship, the *Hot Rod,* using chemical explosives instead of nuclear bombs, and got the craft to lift off the ground.[38] But in 1959, the project fell into the hands of the Air Force, who promptly stopped the flight tests and began planning how to use the concept for military ends. General Thomas Power of the Strategic Air Command proclaimed: "Whoever builds Orion will *control the earth!*"[39] The Air Force envisioned a space-battleship armed with atomic bombs for shooting down enemy missiles which could resist a nuclear explosion 500 feet away by turning its pusher plate toward the explosion and absorbing the shock.[40] Project Orion was abandoned before any serious attempt was made to do an atomic-bomb-powered test flight.

Other *peaceful* uses of nuclear bombs were planned by those involved in *Project Plowshare,* which was based on the concept of using nuclear bombs to create waterways and to mine minerals. In his 1971 book *Man and the Atom,* Glenn Seaborg, AEC chairman in the 1960s, wrote that peaceful nuclear explosives would give humanity a chance to fix up what he called a *slightly flawed planet:* "All of humanity's efforts to restore the *Garden of Eden* have been futile so far. Man's machines have not been powerful enough to compete with the forces of nature."[41]

Edward Teller presented Project Plowshare to the world at the Second International Conference on the Peaceful Uses of Atomic Energy, held in Geneva in 1958. Proposals for *planetary engineering*[42] projects (also referred to as *geographical engineering*)[43] included blasting out *"instant harbors"*[44] all over the world; digging a sea-level canal across Panama; creating a network of waterways linking the Atlantic, the Gulf of Mexico, and the Pacific; and closing the Straits of Gibraltar, which would supposedly cause the Mediterranean to rise and freshen to the point that it could be used to irrigate the Sahara. (Seaborg observed that "of course, the advantages of a *verdant Sahara* would have to be weighed against the *loss of Venice* and other sealevel cities.")[45]

In a 1968 speech, Plowshare enthusiast and JCAE member Craig Hosmer called Plowshare "the *pot of gold,*" saying that this *"newly in-*

vented pump" would make it possible to extract $14 billion of gold, $50 billion of gas, and $1,000 billion of oil, shale oil, and copper, silver, molybdenum, and other metals. Hosmer warned gold speculators (gold was then selling for $35 an ounce) that Plowshare would produce such a gold glut that the price would drop.[46]

The Plowshare program suffered from a number of problems. First, many people questioned whether the intent of the program was purely peaceful. Data gathered from *peaceful nuclear tests* could be put to military use. For example, studying the cratering effects of underground explosions might be useful in calculating how to *bust* missile silos.

Representative Hosmer complained frequently about people who questioned the purpose of the program: "Plowshare R&D test shots have been needlessly cancelled or delayed by '*Nervous Nellies*' in the State Department who are plagued with unfounded fears that a peaceful nuclear shot might shake some disarmament conference to pieces."[47] Hosmer suggested that people "adopt *mature attitudes* about peaceful nuclear explosives."[48]

The Plowshare program was also constrained by two words in the 1963 Limited Test Ban Treaty that ended atmospheric weapons testing by the superpowers. The treaty prohibited nuclear explosions which resulted in "radioactive debris" being detected outside of the country setting off the bomb. Chafing under this restriction, Seaborg proposed that the word *debris* should be interpreted "with the proviso that the 'radioactive debris' . . . would be present in no more than *de minimis* (that is, very minute) quantities."[49]

A third problem for Plowshare proponents was the question of what AEC Chairman James Schlesinger termed "*environmental aesthetics.*" Describing the AEC's experiments using peaceful nuclear explosives to stimulate natural gas production, Schlesinger said the program was "*economically attractive*" and "*technically attractive.*" But there were "some questions with regard to the *environmental aesthetics* of the program, if I can put it that way. A production program of this sort would require a considerable number of shots—perhaps 100 to 200 a year—to have a meaningful program. Whether that is something that the public would welcome at this time is an open question."[50]

Plowshare never produced any "instant harbors," and it has not become a "pot of gold." But it has had one important effect: it provided a justification for the continued development of nuclear weapons. As Fred C. Ilke, the head of the U.S. Arms Control and Disarmament Agency, wrote in 1977, Plowshare

> made it easier . . . for the Indians to pretend their nuclear explosion served only peaceful purposes. People think of swords that have been forged into plowshares as harmless because they cannot be used for war; but so-called "peaceful" nuclear explosions can destroy a city or other targets. Indeed, a basic weakness of some of the past policies has been the assumption that an

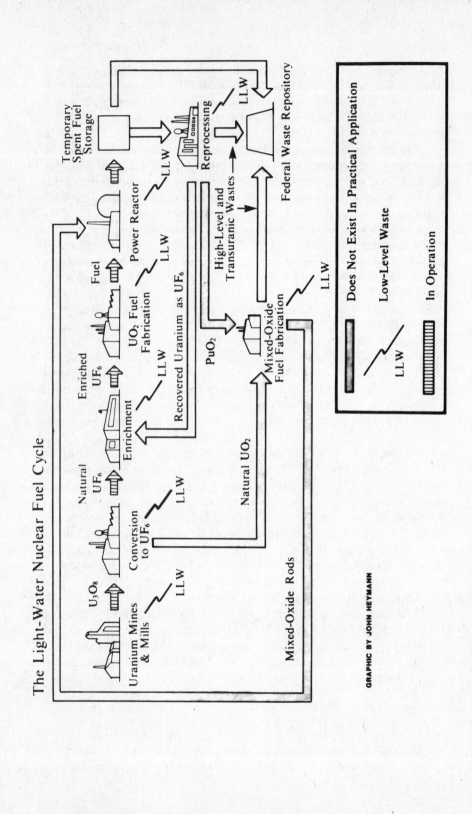

The Light-Water Nuclear Fuel Cycle

GRAPHIC BY JOHN HEYMANN

application of nuclear energy will not be of military use simply because it has some civilian use.[51]

To date, a few of the peaceful nuclear visions of the fifties and sixties have been realized. Radioisotopes, for example, have proved to be very useful in science and medicine. But the more extravagant proposals, like nuclear-powered planes and spaceships, floundered. In 1980, nuclear reactors provided about 11% of the United States' electricity, or only about 3½% of its total energy supply. And, of course, power from nuclear reactors is hardly *too cheap to meter*; indeed, it is costly and becoming more so. Unlimited material abundance is still only a dream. The earth has not been restored to a Garden of Eden. And in the name of Atoms for Peace, the raw materials and technical skills needed for making bombs have proliferated.

FIGURE 5   THE NUCLEAR FUEL CYCLE   When most people think of nuclear power, they think of nuclear reactors—large, centralized industrial facilities that generate electricity as they *"burn"* fissionable material, or *nuclear fuel.* Reactors are only part of a vast, widespread industrial pipeline that has been termed the *nuclear fuel cycle.* The cycle includes everything from extracting raw materials from the ground to disposing of the highly radioactive *waste products* produced in the reactors. The fuel cycle in theory is split into two parts: a *front end,* consisting of those steps that occur before the material reaches the reactor, and a *back end,* or those steps that are supposed to happen after it leaves the reactor. While there are problems and hazards associated with nearly every step of the cycle, it is the back end that has caused the most controversy, since it has never been implemented.

The nuclear fuel cycle begins at a uranium mine, when raw ore is extracted. Later the ore is milled to produce uranium dioxide (U_3O_8), also known as *yellowcake.* Milling also produces huge quantities of spent ore, or *uranium mill tailings,* which emit radioactive *radon gas.* In the western United States where uranium is mined, there are more than 140 million tons of tailings in exposed piles.*

Less than 1% of natural uranium is in the form of the isotope U-235, which can be used as a fissionable material in power reactors. 99.3% of natural uranium is the heavier isotope U-238, which cannot be used as a fissionable material in reactors. Before natural uranium can be "burned" in a light-water reactor—the type most commonly used in the U.S.—or exploded in a bomb, it must be enriched. In this process U-235 is selected out and concentrated. The yellowcake is chemically converted into uranium-hexafluoride (UF_6) and then enriched. Enrichment plants— isotope separation plants such as the ones built by the Manhattan Project at Oak Ridge—exploit tiny differences in weight between the isotopes in order to separate them. The process is difficult and expensive. Natural uranium must be enriched to between 2% and 4% U-235 before it can be used in light-water power reactors, and to about 90% before it can be exploded in a bomb.

Enriched uranium is shipped to *fuel fabrication plants* where it is converted into nuclear *fuel rods* for use in reactors. There are currently seventy-two power reactors licensed to operate commercially in the U.S. These reactors provide about 11% of the nation's electricity, or about 3½% of its total energy supply. The U.S. commer-

*Ronnie D. Lipschutz, *Radioactive Waste: Politics, Technology, and Risk* (Cambridge, Massachusetts: Ballinger Publishing Company, 1980), p. 135.

cial nuclear power industry is based on light-water reactor designs (LWRs)—reactors that are cooled with ordinary water.

In a nuclear reactor, the uranium in the fuel rods undergoes fission, producing enormous quantities of heat that is converted into electricity. During fission, neutrons strike fissionable atoms—mostly atoms of U-235—breaking them apart into smaller atoms, called *fission products*. This process releases energy and more neutrons, which strike more atoms, splitting them, and sustaining a nuclear chain reaction. Some of the neutrons are absorbed by the U-238 in the rods, producing plutonium-239, which is also fissionable. Some of this plutonium in turn fissions. The fission products include other extremely radioactive isotopes such as strontium-90, iodine-131, and cesium-137.

After being "burned" in a nuclear reactor, the *spent fuel*, which contains radioactive fission products and plutonium, continues to generate heat. Spent fuel rods are removed from a reactor and stored underwater in a *spent fuel pool* to prevent them from melting and dispersing their radioactivity.

The spent fuel pool is where the nuclear fuel cycle now stops. Spent fuel from commercial reactors is accumulating in *temporary storage* in spent fuel pools on reactor sites. The final shape of the nuclear fuel cycle is a matter of intense scientific and political debate.

There are now three proposed shapes for the back end of the nuclear fuel cycle. Since it was recognized early that U-235 was too scarce to power the ascent to Eden, nuclear developers proposed to *reprocess* spent fuel rods to extend the life of the fuel supply. Reprocessing would enable them to recover the unfissioned plutonium and uranium still inside the spent fuel rods. Under this scheme, spent fuel would be shipped from the spent fuel pools to a nuclear *reprocessing plant*, where the plutonium would be chemically separated from the fission products, as is done when plutonium is manufactured for weapons. The plutonium would be shipped to a fuel fabrication plant, where mixed-oxide fuel rods, containing plutonium dioxide (PuO_2) and uranium dioxide (UO_2), would be manufactured. The mixed-oxide rods could be used in contemporary power reactors.

The second proposed shape for the back end of the fuel cycle involves the use of *breeder reactors*. Breeders are of a different design than LWRs and would in effect create a new kind of fuel cycle. They are designed to produce more fuel than they consume by converting large amounts of U-238 into plutonium while generating electricity at the same time. The plutonium thus produced could be used in mixed-oxide fuel rods for contemporary plants, or it could be used directly in breeder reactors. Since U-238 is abundant, this would provide an energy source that in theory at least would be virtually inexhaustible. Nuclear developers envision *burning the rocks*—crushing the granite in the earth's crust to extract small amounts of U-238 for breeders. As Glenn Seaborg, a former AEC chairman, and William R. Corliss wrote in 1971, burning the rocks would help prevent human civilization from being "just a brief fossil-fuel flicker in the long cosmic night."**

Technical, political, and economic problems with both breeder reactors and reprocessing have led some nuclear developers to propose a third back end for the fuel cycle, namely, that spent fuel rods be directly disposed of in *permanent waste repositories*. Such repositories would be located either underground or in the ocean floor.

No matter what shape is finally selected for the nuclear fuel cycle, the final step must be the disposal of radioactive waste. The highly radioactive fission products that are made in reactors must be isolated from the environment for thousands of years, and no method or facility for accomplishing this has yet been developed.

**Glenn T. Seaborg and William R. Corliss, *Man and Atom* (New York: E. P. Dutton and Co., Inc., 1971), p. 37.

PART TWO

The Ministry of Truth

Don't be dissuaded by the pessimists who say nothing can be done. Start with an education program—make sure everyone understands the potential damage in the exposure of sensitive intelligence sources and methods. Check your distribution lists—do all those people really have need-to-know? How about your information officers? In their zeal to publicize your mission, your capabilities, your product, are they giving away the family jewels faster than you can polish them? Are you sure that it's always in that other place where the leak took place? And a very sticky high level one—when someone decides to release some information for what he believes to be valid political purposes:

—are the procedures crystal clear?
—does he have the authority to do it?
—has the material been officially declassified or at least sanitized?
—has he consulted with the appropriate Senior Intelligence Officer?

You might say we're tilting at windmills with this one—perhaps so. I say let's tilt. And let's get the law to back us up! It's our national security!

Mike Levin
National Security Agency
Speech given at the 15th annual meeting of the
National Classification Management Society, 1979[1]

The nuclear controversy involves people's hearts more than their heads. The public isn't won over by facts and statistics. The nuclear debate isn't over whose facts are correct, but, instead, who can come up with the greater hazard and have it successfully perceived so by the people. So forget the facts once in a while. Counter the activists not with facts but with closed factory gates, empty schools, cold and dark homes and sad children. . . . Once the emotional chord is struck, the sound will carry to the state and Federal political arenas where the outcome of the nuclear controversy is being decided.

Frank B. Shants
Public Service Company of New Hampshire, 1978[2]

CHAPTER FIVE

The Secrecy System

The power of the Department of Energy to restrict the flow of information is awesome. Whatever the DOE declares secret remains secret until the DOE determines it no longer is secret. The documents that stipulate which information is secret are secret. And the very existence of a particular secret document may itself be secret. Moreover, the DOE has the legal authority to prevent any information having anything to do with nuclear technology—whether related to atomic weapons or power production—from being made public.

All of this is done in the interest of protecting *national security*, a vague concept officially defined as "the national defense and foreign relations of the United States."[3] This expansive definition allows the government to conceal much information that should be made public in any nation purporting to be a democracy.

The government officially denies that it uses classification to restrict debate. Interviewed in 1980, Murray L. Nash, deputy director of the DOE's Office of Classification, said that "there's not even the slightest sliver of truth" to this charge. "Everything we've kept secret, we've kept secret for *sound technical reasons*, without any political implications whatsoever."[4]

Nash's claim has to be viewed with skepticism. In this book we present many examples of the use of official secrecy to restrict debate, to shield projects from public scrutiny, and to cover up mistakes:

- ☐ the use of censorship to limit public debate about whether to build the hydrogen bomb (chapter 5)
- ☐ the suppression of information during the fallout debate of the 1950s (chapter 7)
- ☐ the coverup of the nuclear accident that took place in the Ural Mountains of the U.S.S.R. (chapter 9)
- ☐ the use of classification to conceal radioactive waste problems at our government's Hanford reservation (chapter 13)
- ☐ the refusal of the government to discuss the implications of false nuclear alerts that could lead to accidental nuclear war (chapter 20)

In this chapter, we examine the nuclear secrecy system—the way it functions and what it accomplishes.

THE CLASSIFICATION PRIESTHOOD

In 1972, Alvin Weinberg, director of the Oak Ridge National Laboratory from 1955 to 1973, wrote an article in the journal *Science* in which he suggested that a *technological "priesthood"* might be necessary to ensure that nuclear power was managed properly in perpetuity. Weinberg argued that nuclear technology is a particularly demanding technology, and that society might need to evolve new social institutions to handle it. The development of nuclear weapons, he wrote, "in a sense . . . established a *military priesthood* which guards against inadvertent use of nuclear weapons, which maintains what a priori seems to be a precarious balance between readiness to go to war and vigilance against human errors that would precipitate war. . . ." Similarly, "peaceful nuclear energy probably will make demands of the same sort on our society, and possibly of even longer duration."[5]

In 1973, Weinberg explained his vision of a technological priesthood more fully:

> [N]o government has lasted continuously for 1,000 years: only the Catholic Church has survived more or less continuously for 2,000 years or so. Our commitment to nuclear energy is assumed to last in perpetuity—can we think of a national entity that possesses the resiliency to remain alive for even a single half-life of plutonium-239? A *permanent cadre of experts* that will retain its continuity over immensely long times hardly seems feasible if the cadre is a national body. . . . The Catholic Church is the best example of what I have in mind: a central authority that proclaims and to a degree enforces doctrine, maintains its own long-term social stability, and has connections to every country's own Catholic Church.[6]

Like the technological priesthood foreseen by Weinberg, the national classification elite is a kind of secret society, closed to the uninitiated. It is a sect marked by rigorous internal discipline, highly developed rituals, a strict hierarchy, and a consistent philosophy. Central to this philosophy is the principle of *compartmentalization*, which holds that the best way to control information is to break it into small pieces, and never to allow too much to be assembled in one place. No person should be trusted with more information than he or she *needs to know* for the performance of official duties. No one outside the order should be trusted with secret information at all. Any inquiry into whether the government is misusing its powers to classify quickly runs up against a stone wall.

The classification priesthood has developed an elaborate system to protect its secrets. The priesthood makes a distinction between classifying documents and classifying the information contained within them. If a piece of *information*, broadly defined to include facts, concepts, ideas, and knowledge, is classified, then all documents that contain the information need to be classified. The term *document* is defined as "any record of information regardless of physical form or characteristics."[7]

The DOE keeps a staggering amount of information secret. The number of DOE documents that are classified reaches into the millions. While the department boasts that two and one-half million documents have been declassified since 1946, it will not provide an estimate of the number of documents that remain classified. Claiming that he did not know the number of classified DOE documents, Nash of the Office of Classification would say only that "probably more" than two and one-half million are still secret.[8]

In this age of computers, cameras, and copying machines, maintaining tight control over large quantities of information and huge numbers of documents is no small chore. The classification priesthood has developed an extensive set of regulations, which is administered by the security forces attached to the agencies that handle classified information. The basic strategy is simple: lock sensitive information away—behind closed doors or in trustworthy minds.

The most important step in proper compartmentalization is to restrict the membership of the priesthood. Before being admitted, each applicant must be *cleared*, that is, granted a *security clearance*. A security clearance is bestowed only after a thorough investigation of the applicant's background has been completed by the Civil Service Commission of the Federal Bureau of Investigation. If the investigation turns up any *derogatory information* which the applicant is unable to rebut, the clearance normally will be denied.

By accepting a security clearance, an individual officially becomes a member of the classification priesthood, agreeing to participate in its rituals and to abide by its regulations. Above all, the person must swear never to disclose classified information to *unauthorized persons*, even years after his or her security clearance has been terminated. Before they are allowed access to any secret information, new DOE employees and contractor employees are given an *Initial Classification Indoctrination*.[9] The DOE maintains a program of *continuing classification education* to make sure that no one forgets his or her responsibilities.[10] (See figure 6.)

Once a person accepts a security clearance, his or her ability to engage in public debate is compromised. In the 1950s, for example, the government used its power over physicists with security clearances to make it difficult for the American people to hear a well-informed debate on the effects of radiation on health. When the Washington chapter of the Federation of American Scientists sponsored an open meeting on Nuclear Radiation Hazards and their Political Implications in April, 1955, the planning committee had a hard time finding a physicist to appear on the program.

M. Stanley Livingston, professor of physics at the Massachusetts Institute of Technology, described the committee's difficulties:

> Refusals [were made for] . . . various reasons. Some described the impossibility of speaking authoritatively on a topic in which so much information is still classified; others expressed a downright fear of involvement in a controversial issue which might cause them *security clearance trouble.* After many refusals the planning committee did obtain acceptances from two physicists who are employed in the Defense Department. However, these physicists later withdrew when the Defense Department imposed new restrictions on public statements by its employees. . . ."

Summing up the frustrating search, Livingston said:

> This experience illustrates one of the political dilemmas in which we find ourselves. . . . Those who know won't speak, and those who don't know cannot speak with authority.[11]

The *need-to-know* policy limits the flow of information even within the priesthood. This policy states that people should be allowed access only to that information which "is essential to the accomplishment of official Government duties or contractual obligations."[12] Need-to-know must be meticulously demonstrated—even by those with security clearances. As the DOE's Nash said, "Curiosity doesn't qualify."[13]

The need-to-know policy applies to all classified information, including *classification guides*, the books which spell out what information is classified, what category it falls into, and how *sensitive* it is. With one exception, the *Guide to Unclassified Fields of Research*, all of the DOE's classification guides are themselves classified.[14] Most members of the priesthood do not need to know what is in the guides; only people who are specifically authorized to use them are permitted to view those sacred texts.

Because the classification guides are classified, it is hard for the public to detect *unofficial classification*, the technique of saying something is classified when actually it is not. As one government employee confided in a 1980 interview, "It used to be that if a journalist asked a question you didn't want to answer, you'd tell him that the matter was classified, when lots of times you knew it really wasn't. Now we don't do that so much anymore."[15]

Information that is officially classified falls into one of three categories: *National Security Information, Restricted Data,* and *Formerly Restricted Data,* also known as *NSI, RD* and *FRD.* National Security Information is the kind of classified information normally handled by federal agencies. Restricted Data and Formerly Restricted Data are special categories that have to do with atomic energy.

Under the current definition of *National Security Information,* information of the following kinds may be classified as *NSI* and declared secret, if its unauthorized disclosure "could be *expected* to cause at least *identifiable*

CLASSIFY IT NOW

FIGURE 6 Poster produced by the Energy Research and Development Administration (now subsumed under the Department of Energy) to remind agency employees and contractors of their security responsibilities.

damage to the national security": information about military plans, weapons or operations, about foreign governments, about intelligence sources or methods, about scientific, economic or technological matters related to national security, about programs for safeguarding nuclear materials and facilities, or other categories of information related to national security.[16] Determination of what "could be *expected*" and what constitutes "*identifiable damage*" is vested in high officials of the classification elite called *original classification authorities*.

Original classification authority, the power to classify information which has not been classified previously, has been given to the heads of

National Security Information can be classified by the heads of any of the following federal agencies, or by subordinates or agency contractors who have been delegated classification authority:

Department of State
Department of Defense
Department of Energy
Department of the Treasury
Department of Commerce
Department of Transportation
Department of the Army
Department of the Navy
Department of the Air Force
Nuclear Regulatory Commission
Department of Justice
Arms Control and Disarmament Agency
Central Intelligence Agency
Agency for International Development
International Communication Agency
National Aeronautics and Space Administration
General Services Administration
Overseas Private Investment Corporation
Export-Import Bank

Source: Executive Order 12065.

nineteen federal agencies and eight officials within the Executive Office of the President (see figure 7). These people can delegate original classification authority to subordinates and government contractors.[17]

The Atomic Energy Act of 1946, the law which created the Atomic Energy Commission, also created the special kind of classified information called *Restricted Data (RD)*. Any information related to the design, manufacture, or utilization of atomic weapons, the production of enriched uranium or plutonium, or the use of those materials for the production of energy is Restricted Data, unless the information has been removed from the Restricted Data category by order of the Secretary of Energy (or formerly, by the head of the Energy Research and Development Administration or the Atomic Energy Commission).[18]

Unlike any other kind of classified information, Restricted Data is *born classified*. This means that any information, any idea, any concept that falls into the RD category is classified—"no matter where originated or by whom generated"[19]—the instant it is conceived. It remains classified unless the DOE *declassifies* it; no other agency has the power to declassify Restricted Data. No action need be taken to classify Restricted Data; it is automatically classified by law. This procedure is directly opposite to the procedures for classifying National Security Information. While it takes a positive action to classify NSI, it takes a positive action to declassify RD. It makes no difference whether or not the person who develops Restricted Data is a government employee or contractor.

The born classified doctrine has caused some problems for people who like to think about nuclear topics, and has forced them to watch closely what they say. In an article in the *New York Times Magazine* in 1956, nuclear engineer V. Lawrence Parsegian gave some ironic examples of the problems that *classification at birth* poses:

> On one project, several scientists who do have . . . [security] clearances and some knowledge of classified work are associated with a Professor X who does not have a clearance; the reason is, his citizenship has not yet been granted. However, Professor X, a renowned scientist, has an embarrassing faculty for producing ideas that fall in the classified realm; in effect, he thinks *classified thoughts*! His colleagues cannot discuss his own ideas with him—not because they aren't cleared, but because he isn't.

Parsegian went on to say that "even in lecturing before graduate practicing engineers the writer has frequently had to confine his remarks to reading from card notes, lest he say more than he should."[20]

The born classified doctrine makes it hard to investigate the government's nuclear policy. According to Howard Morland, author of an article on the H-bomb that triggered the legal battle known as *The Progressive* case (discussed later in this chapter), the keepers of the secrets "do not simply withhold the answers; *they can also confiscate the questions*" (original emphasis).[21]

Morland's comment was based on an experience he had in 1978. At that time, the government was expecting a shortage of plutonium for weapons manufacture. The DOE, which manufactures nuclear weapons, was considering starting up a standby plutonium production reactor at its Savannah River Plant in South Carolina and rebuilding its old plutonium reprocessing plant in Hanford, Washington. Morland wanted to know why the DOE was expecting the shortage. Would the neutron bomb require more plutonium than ordinary tactical nuclear weapons? Was the extra plutonium needed for new warheads for the Trident submarine and the proposed MX missile? Was the whole plan just a pork-barrel project for a few politicians' home districts?

In an effort to get a public answer to these kinds of questions, Morland enlisted the aid of Representative Ronald Dellums of California. Dellums sent a letter to the DOE saying that a constituent had requested answers to some questions about the federal plutonium production program. Dellums explained that the information was not for his use, nor for the use of the House Armed Services Committee, of which he was a member, but for a member of the public. And he pointed out that the reconstruction and reopening of plutonium production plants would have substantial financial and environmental costs. Dellums wrote that the "American people need to know the reasons for the anticipated plutonium shortage in order to have informed opinions on the cost-benefit aspects of the plutonium production issue." Attached to the letter was Morland's list of about a dozen questions.

In three weeks, Dellums received the DOE's reply. Saying that it "is not possible to respond to most of the questions in an unclassified manner," the DOE declined to answer all but one. More chilling was the following sentence: "The enclosure to your referenced letter contains 'secret/restricted data' and should be so classified." The enclosure was Morland's list of questions. It had been made secret.[22]

The third category of classified information is called *Formerly Restricted Data*. This name, which seems to imply that the information is no longer classified, means nothing of the sort. Formerly Restricted Data is information that has been removed from the Restricted Data category after the DOE and the Department of Defense have jointly determined that it relates to the military utilization of atomic weapons and can be adequately safeguarded as National Security Information. This recategorization allows the Defense Department more flexibility in the security procedures used to handle the information. Formerly Restricted Data is still classified; it has been *transclassified* from one category to another.[23]

According to Nash, about 90% of the classified information that the DOE manages is either RD or FRD. The remaining 10% is NSI—mostly information about foreign governments, international energy resources, military plans, the security of strategic nuclear materials, and intelligence operations.[24]

All classified information, whether RD, FRD or NSI, is assigned a *classification level* which indicates its degree of sensitivity. The more sensitive the information, the more strict the security procedures that are followed when it is handled. Classification levels range from *Confidential* (the least sensitive), through *Secret*, to *Top Secret* (the most sensitive). In addition, according to David Wise, author of *The Politics of Lying*, there is a category of National Security Information called *Special Intelligence*. "The public, and even most members of Congress," Wise wrote, "are generally unaware of the existence of a whole spectrum of super-secret classifications, which are, in effect, above and beyond Top Secret." These designations include *TK*, which refers to information obtained by intelligence satellites, and cryptic words like *DINBAR, TRINE, SPOKE, UMBRA,* and *HARUM*.[25]

The classification priesthood takes great care to ensure that classified information is "used, processed, stored, reproduced, and transmitted only under conditions that will provide adequate protection and prevent access by unauthorized persons."[26] Security measures include guards, alarm systems, special couriers to transport documents, identification badges, and special telecommunications equipment for electronic transmission of classified material. Classified documents can only be stored in government-approved security containers—for example, vaults or steel file cabinets with combination locks. Even *classified waste* must be protected: it is deposited in locked trash cans, where it remains until it is destroyed by burning or shredding. Federal security forces carry out periodic inspections of installations that handle classified information to make sure that security procedures are not being violated.

A further precaution used with very sensitive information is simply to avoid writing it down. The DOE's unclassified *Guidebook for the Authorized Classifier* reminds the reader that it "is DOE policy to keep the number of Top Secret documents as low as possible, to minimize both the amount of extremely sensitive information appearing in writing and the administrative burden of Top Secret handling."[27]

Throughout the *DOE-contractor complex*, which includes the DOE's headquarters and field offices and its contractors, there are 223 authorized original classifiers.[28] They expend a great deal of effort making sure that no documents containing classified information escape classification. Catching all the Restricted Data is particularly troublesome. Because RD is classified at birth, there is the problem of keeping track of new information that is generated outside the government. The DOE keeps an eye on research that it believes might produce classifiable results, claiming that it "has a responsibility under the Atomic Energy Act to *monitor* R&D conducted by private organizations and individuals."[29]

It is a tricky business to decide how much information needs to be classified. Decisions about what to classify are made "on the basis not only of what the information itself reveals, but also on what may be revealed through *association with unclassified information*."[30] Original classification authorities try to foresee the impact of each piece of information, imagining all the ways it could be combined with information from all other sources.

In classifying documents, the basic rule is that every document should be assigned the most restrictive classification that might apply to it. Thus a document that contained some information classified Secret and some classified Top Secret would be assigned a Top Secret classification.

There is a procedure for challenging the decisions of the classification priesthood, but it is limited. Anyone who believes that a specific piece of information or a certain document has been improperly classified can request a *mandatory classification review*. The classification priesthood is the final arbiter of these challenges, however. If the DOE's director of classification rules that the information or document must remain classified, then the requester is notified in writing. The DOE may also provide "a brief statement as to why the information or documents cannot be declassified."[31] When deemed *appropriate*, the DOE may supply a *"sanitized"* version of the requested document, with all classified information removed. An *"ap-*

propriate unclassified description" of the classified information may also be provided.[32] Unsatisfied requesters can appeal the director of classification's rulings by filing a request with the *Classification Review Committee*, an internal DOE committee that sits at the pinnacle of classification authority. The decisions of the Classification Review Committee are final.

Normally, the DOE will acknowledge the existence or nonexistence of a classified document requested by a member of the public. But the priesthood leaves itself a loophole in this general rule. The DOE can—and will—refuse to confirm or deny the existence of documents if "the fact of their existence or nonexistence would itself be classifiable."[33]

DOWN THE MEMORY HOLE

The extraordinary nature of the classification priesthood's claim to be the sole judge of what should be secret was fully revealed in 1979 in a legal battle known as *The Progressive* case. In an effort to prevent *The Progressive*, a small monthly magazine, from publishing an article which the priesthood claimed contained the *secret of the hydrogen bomb*, the government pushed its case to extremes:

- ☐ A federal court imposed *prior restraint*, the legal term for censorship prior to original publication.

- ☐ The government advanced the concept of *retroactive secrecy*, declaring that previously published articles contained secrets.

- ☐ The government argued that *technical information* was exempt from the First Amendment's guarantee of freedom of the press.

- ☐ Finally, the government argued that decisions about classification are *too complex* to be reviewed by the courts, claiming that the actions of the classification priesthood are exempt from judicial review.

The case was sparked by an article called "The H-Bomb Secret: How we got it, why we're telling it," which *The Progressive* planned to publish in its May, 1979 issue. The central point of the article, which was written by free-lance journalist Howard Morland, was that the *"H-bomb secret"* was not a secret. The DOE claimed that about 20% of the text and all of the diagrams accompanying the article contained Secret/Restricted Data—very sensitive information on nuclear weapons design.

In a sense, the roots of the case go back to 1950, when President Truman gave the order to proceed with the construction of the hydrogen bomb. The decision, one of the most momentous of this century, was made without involving the public to any significant degree; secrecy would not permit it. A small group of politicians and scientists took part in the deliberations that led to Truman's decision. As physicist Louis N. Ridenour wrote in *Scientific American* in March, 1950, "A major issue of public policy, one quite possibly involving our national existence, was decided in a fully authoritarian way. Not without public discussion, to be sure, but without anything that could have been called informed public discussion. The public did not even know, and still does not, what the actual questions at issue were."[34]

Scientific American wanted to put the facts about the *superbomb*, as it was then known, before the people. In April, 1950, the presses started to roll with a second H-bomb article, written by Cornell physicist Hans Bethe. Then agents of the Atomic Energy Commission arrived. They stopped the presses, destroyed 3,000 copies of the magazine, and insisted on *rewriting* the article in a *sanitized* form.[35] Publisher Gerard Piel charged that the AEC was "suppressing information" that the American people needed to know in order to form intelligent opinions and make informed judgments about the superbomb. Bethe, who also worked for the government as a consultant, didn't want to fight the AEC.[36] The magazine instead acceded to the AEC's editing, and published what Piel called "a mutilated article."[37]

When the government offered to rewrite Morland's 1979 article in sanitized form, *The Progressive* refused to back down, arguing that the government had no right to censor the article because it contained no secrets. Acting under the authority of the Atomic Energy Act, the Department of Justice filed for a temporary restraining order to prevent the article from being published. The case was heard by Federal District Judge Robert W. Warren, a Nixon appointee. Observing that "you can't speak freely when you're dead," Warren granted the government's request, temporarily prohibiting publication of the article pending a full hearing on the issue.[38]

The case was quickly covered by network television and appeared on the front pages of the nation's newspapers. The title and intent of the Morland article were distorted. The government portrayed Morland and the editors of *The Progressive* as fools or fanatics, who showed no concern for nuclear proliferation and were intent on violating the sanctity of one of the country's most sacred secrets by publishing a do-it-yourself manual on how to build an H-bomb.

Yet the popular image of an *atomic secret* as some diagram or formula on a single sheet of paper has little to do with the real secrets that remain in the world of nuclear weapons design. (According to David Lilienthal, the AEC's first chairman, a newspaper photographer called the AEC Public Relations Department when control of atomic development passed from the Manhattan Project to the AEC in 1946, and asked to take a picture "of General Groves handing *the secret* to Chairman Lilienthal.")[39] As Gerard Piel wrote in 1979, it is not the lack of secret information that stops nations from building hydrogen bombs, but the lack of the industrial and scientific base: "There are various ways to make these weapons and there are wrinkles in the manufacturing that are secrets of the trade-secret kind. But that kind of secret takes a lot of explaining and, nowadays, a carload of computer printouts; it cannot be dropped inadvertently in a discussion of public policy or even in an overcomplicated popular exposition." Moreover, Piel suggested that long before a true secret could wind its way into the news media it would have leaked to the "communications channels that count in the espionage business."[40]

If there was no secret to protect, what was the classification priesthood protecting? Morland believed the answer to that question was that the nation's nuclear policy makers were using "the *illusion* of secrecy to deny information to the public and to exclude citizens from the process of government" (original emphasis).[41] Congressional meetings on nuclear weapons are held in secret. The federal nuclear industry that manufactures, transports, and stockpiles nuclear weapons is exempt from normal environmental review. Policy decisions that will affect the future of civilization are being made without informed public debate.

The Progressive based its defense on evidence that Morland's article did not contain any secrets. As the trial approached, the editors of the magazine frantically searched for scientists who could write affidavits saying that the information in the article was already in the *public domain*. This search was hampered by security regulations. Since the article had been classified, only persons with a *Q clearance*—the kind needed to permit access to Secret and Top Secret Restricted Data—could read and comment on it.

The Progressive finally found a number of scientists who were able and willing to read the article. Their affidavits said that any competent physicist could readily obtain all the ideas and information in Morland's article from unclassified sources. Moreover, they said that the ideas could be obtained not within years, but within hours.

The government responded by classifying the affidavits. The full texts were available only to Judge Warren. Sanitized versions were provided to outsiders. The lawyers representing *The Progressive* were forced to obtain security clearances in order to participate in the legal proceedings. Morland and the editors refused to obtain clearances, because doing so would permanently interfere with their ability to write critically about the government's nuclear policies. As a consequence, they were unable to attend the closed hearings, and a wall of secrecy was built between the defendants and their lawyers.[42]

In addition to classifying the affidavits, the government advanced the concept of *retroactive classification*. An article written by Hans Bethe which had been published in an eighth-grade encyclopedia was declared retroactively secret. Morland's copy of his college physics book was declared secret because he had underlined portions of it. The Justice Department announced that arguments over whether an encyclopedia article or an old textbook actually contained secrets were *secret arguments* that could only be conducted in secret.[43]

On the basis of secret arguments conducted during closed hearings, Judge Warren ruled that the magazine could not publish Morland's article. Warren summarized his reasoning in the terms of a crude risk-benefit analysis: a mistake in ruling against *The Progressive* would reduce First Amendment rights, while a mistake in ruling against the government "could pave the way for thermonuclear annihilation for us all."[44]

Many were not convinced that the judge's reasoning was sound. In its editorials, the *New York Times* called the government's case "lame in both logic and law," pointing out that any "secret" easily penetrated by a "self-trained student of the nuclear weapons program" could not be much of a secret. The newspaper charged that government witnesses clouded the air with "technical jargon and dark forebodings about the fate of all mankind."[45] Calling Judge Warren's decision "unconstitutional, unnecessary, arbitrary and ill-informed," the *Times* said that the judge had become "just one more victim of a system of secrecy that permits the cry of 'national security' to overwhelm our most precious safeguards against suppression of speech."[46]

The Progressive decided to appeal Judge Warren's ruling. As the appeal date approached, the government moved onto increasingly shaky ground as it struggled to contain a secret that was not secret. The DOE suddenly closed the library at the Los Alamos Scientific Laboratory when a researcher working for the attorneys for *The Progressive* began examining unclassified documents on nuclear weapons design. When the researcher returned from lunch, one of the documents he had been using was missing, all references to it had disappeared from the card catalog, and the librarian claimed to know nothing about the incident. Shortly thereafter, the DOE's *security division* arrived, and the library was closed for *"inventory and review."* Later, the government claimed that the documents had been *"improperly classified."*[47] The LASL library was shifted into *classified mode*, a term that, in plain English, meant that it would be closed to the public for an indefinite period.

Hopeful that the revelations about the Los Alamos library incident would force Judge Warren to recognize that the H-bomb secret was out, the lawyers for *The Progressive* asked Warren to end the ban on publication. Warren rejected their motion. Moreover, he kept the reason for this decision secret, withholding the text of his statement.[48]

During the appeal, the government advanced bizarre theories to support its position. It claimed that *technical information* was exempt from the provisions of the First Amendment, arguing that scientific knowledge plays no vital role in political discussion,[49] an insupportable claim in our technological society.

As another line of attack, the government argued that any decision as to whether a piece of information could safely be declassified was *too complex* and technical to be left to the courts.[50] Only properly trained classification experts were competent to make classification judgments, and the courts should not interfere with these judgments. If this proposition had been accepted, classification decisions would now be exempt from judicial review.[51]

A verdict was never reached in the appeal because the government suddenly dropped its case against the magazine. This turnaround was prompted by the publication of a letter which the government said also contained the H-bomb secret. The letter, written by a California computer programmer and amateur nuclear weapons designer named Charles Hansen, described the operation of a thermonuclear bomb. It was accompanied by a diagram drawn with the aid of a tuna fish can and some jar lids.[52] After Hansen mailed copies of the letter to the DOE and a half-dozen newspapers, the DOE quickly classified it, saying it contained Secret/Restricted Data. But the government was unable to retrieve its "secret." Several newspapers refused to turn over the letter, and copies proliferated. When the *Press Connection* of Madison, Wisconsin published the letter, and the *Chicago Tribune* informed the DOE that it planned to do the same, the government dropped its case against *The Progressive*, explaining that after publication of the Hansen letter it would be "pointless" to keep Morland's article secret.[53] After six months of being under prior restraint, *The Progressive* was finally free to publish Morland's article.

The victory for *The Progressive* dealt a blow to official secrecy, but it left the legal basis for suppression of information and prior restraint unscathed. No ruling was made on the government's claim that classification decisions are exempt from judicial review, nor on its claim that technical information is not covered by the First Amendment. A prior restraint was allowed to stand for six months and was not ruled unconstitutional. The Department of Justice was quick to point out that the secrecy provisions of the Atomic Energy Act remained intact.

A number of disconcerting lessons can be learned from the case. For one thing, it demonstrated that anyone who is serious about asserting his or her First Amendment rights in defiance of government censorship will end up paying very dearly. *The Progressive*'s legal fees exceeded $200,000.[54] The Justice Department paid its legal fees with taxpayers' money, and no records were kept of the time its lawyers spent prosecuting *The Progressive*.[55]

The case also underlined the extent to which the American media engage in self-censorship. In its May/June, 1979 issue, the *Columbia Journalism Review* noted that since the beginning of the nuclear age, there had been a "silent understanding" between the government and the press that journalists would accommodate themselves to nuclear secrecy. Because it refused to submit Morland's article to the government's editing, "*The Progressive* . . . found itself challenging an established consensus—the belief that, so far as nuclear weaponry is concerned, the public doesn't *need to know* what the press doesn't want to know."[56]

The main lesson the government learned from the case was that it needed to tighten security. As the *New York Times* reported, "[M]any military analysts believe that the United States has been too lax and that stricter censorship of magazines, newspapers, reference books and student texts is required."[57]

Some believe that a lesson of *The Progressive* case is that it may not be possible to have a free press and a free people in a nuclear society. As Laurence H. Tribe, a Harvard law professor and a well-known specialist in constitutional law, and David H. Remes wrote in the *Bulletin of Atomic Scientists*, the greatest threat to liberty in *The Progressive* case

> came not from the government or Judge Warren—although both may be faulted for their insensitivity to constitutional concerns—but from nuclear power itself. As long as the commercial and military use of nuclear power is internationally sanctioned, governments cannot be denied broad powers to protect their citizens from annihilation by nuclear war or devastation by nuclear accident.[58]

Tribe and Remes warned that "the erosion of First Amendment rights may be an unavoidable *fallout* of life in the atomic age." The restrictions on First Amendment freedoms could spread from the nuclear arena, and would "threaten to engulf free-speech rights across the board."[59]

CHAPTER SIX

Polls, Ads, and PR

Nuclear developers have always shown great concern for the public-relations impact of their actions. The government and the nuclear industry have conducted an ongoing PR campaign to present the *sunny side of the atom* to the public, both within the U.S. and in other countries. This campaign has played an important role in spreading and defending the nuclear mindset.

Public relations was an important concern during Operation Crossroads, 1946 bomb tests at which the world press viewed nuclear explosions for the first time. Operation Crossroads, a joint Army-Navy-Air Force undertaking, was conducted by Joint Task Force One, created by President Truman in January, 1946. The two explosions were carried out at Bikini, a Pacific atoll roughly 2,000 miles southwest of Hawaii.

Joint Task Force One was an enormous venture. It comprised 41,963 men, 37 women, 242 ships, 156 airplanes, 4 television transmitters, 750 cameras, 5,000 pressure gauges, 25,000 radiation recorders, 204 goats, 200 pigs, 5,000 rats, and the atomic bombs. Every day this armada of men and 37 women consumed 70,000 candy bars, 40,000 pounds of meat, 89,000 pounds of vegetables, 4,000 pounds of coffee, and 38,000 pounds of fruit.[1]

The press and public-relations effort was also mammoth. A special cancellation for mail sent from Bikini on *"atom bomb day"* was prepared, reading "Atom Bomb Test, Bikini."[2] At the first explosion, there were 114 U.S. news correspondents, plus observers and press from several foreign countries.[3] According to *Bombs at Bikini*, the official history of Operation Crossroads, the public-information officers chose from a wide variety of news agencies when selecting correspondents, resulting in the publication of accounts "in nearly every type of newspaper and magazine, including even some of our most 'feminine' magazines."[4]

The press was headquartered on the *Appalachian*, with older correspondents getting the best quarters. On board this ship, there were lectures, films, and a variety of press packets, all designed, as *Bombs at Bikini* put it, to make Operation Crossroads "the *best*-reported as well as the *most-*

reported technical experiment of all time" (original emphasis).[5] The public-relations specialists were especially eager to dispel rumors that the test would cause underwater landslides, create giant tidal waves, rip the crust of the earth, or even set off a chain reaction in water and cause the ocean to burn or explode.

In the first test, *Test Able*, an atomic bomb was detonated in the air over a target fleet which included nine battleships, two aircraft carriers, twelve destroyers, eight submarines, and an assortment of landing craft, drydocks, barges, and merchant ships. The fleet was assembled from captured German and Japanese vessels and old U.S. ships.[6] It was used as a platform for exposing pigs, goats, rats, and an array of military odds and ends to the blast. These items included tanks, weapons carriers, trucks, tractors, airplanes, guns, mortars, rocket launchers, rifles, torpedoes, mines, depth charges, bombs, fuses, rockets, flares, telescopes, gas masks, grease, gasoline, canned apples, apricots, tomato juice, string beans, creamed corn, bacon, turkey, butter, jackets, trousers, parkas, undershirts, drawers, socks, boots, helmets, DDT, and skis.[7]

The biggest public-relations problem at the first test turned out to be the experimental animals placed on board the target fleet to test the effects of radiation. Their fate touched the public, and more than half of the letters of protest condemned the use of animals for this purpose.[8] In California, the San Fernando Valley Goat Association planned a memorial service for the goats killed at Bikini, complete with a bugler playing taps and the lowering of the American flag. Veterans' groups protested the playing of taps and the use of the flag, however, forcing the association to substitute a moment of silence instead.[9] *Bombs at Bikini* assured the public "that radiation sickness is *essentially painless*"—a blatant falsehood. The book went on to explain that "in the case of animals, victims have no mental anguish such as would presumably assail human beings. The animal languishes and either recovers or dies a *painless death*. Suffering among the animals as a whole was *negligible*."[10]

The media's favorite animal, according to *Life* magazine, was *Pig 311*, "the little animal that defied the big explosion."[11] The unfortunate piglet had been locked up inside the Japanese cruiser *Sakawa*, which sank after the first test. Somehow Pig 311 escaped from certain death and was found swimming around in the lagoon. Pig 311 returned home to a heroine's welcome in the U.S. Organizers of the Texas "Air Day" celebration in Harlingen, Texas requested that Pig 311 appear as a "distinguished guest" for the two-day event. Admiral Nimitz sent regrets on behalf of the famous piglet.[12] In a report on the Bikini tests a year later, *Life* featured a full-page photo of Pig 311 in her quarters at the Naval Medical Research Institute at Bethesda, Maryland. Pig 311 had grown to be a magnificent 350-pound sow. *Life* reported ruefully that all attempts to breed her had been unsuccessful, indicating that the atomic explosion "may have made Pig 311 sterile."[13]

The second test, *Test Baker*, was an underwater explosion. It produced a much more serious and long-lived public-relations problem than the experiments on the animals: radioactive fallout. When nuclear weapons are detonated high in the air, little *close-in fallout* is produced, since the radioactive fission products are carried high up into the atmosphere. But weapons detonated near the ground or water produce large amounts of close-in fallout, since dust or water droplets carry the radioactive materials produced by the explosion back to the earth.

The high levels of radioactive contamination produced by the underwater explosion at Bikini were not anticipated. Ninety percent of the target fleet was contaminated with radioactive materials.[14] David Bradley, a young doctor who served as a *Geiger man* at Bikini,[15] described the hazards of fallout from this test in his book *No Place to Hide*. Bradley explained how reef fish, when sliced, dried, and laid on photographic plates, produced radioautographs, photos of their internal organs made by the radioactive elements the fish had taken up from the sea.[16] Ships that entered the lagoon after the test had their saltwater lines and evaporators contaminated by radioactive water drawn in from the lagoon.[17] (Fallout from weapons tests remained a troublesome PR problem for nuclear developers for many years; the AEC devoted a considerable amount of effort to downplaying its hazards—see chapter 7.)

Beginning in the mid-1950s, the AEC conducted a huge public-relations operation to promote the vision of *Atoms for Peace*. This ideological support for nuclear development was important in getting the U.S. commercial nuclear industry started. AEC public-relations activities also helped sell *peaceful nuclear programs* in other nations. The AEC used a wide range of PR techniques, including films, brochures, TV, radio, nuclear science fairs, public speakers, traveling exhibits, and classroom demonstrations. The agency set up the American Museum of Atomic Energy in Oak Ridge, Tennessee, which has been visited by millions of people. The AEC helped the Boy Scouts to establish an *Atomic Energy Merit Badge*. AEC Chairman Glenn Seaborg personally presented the first merit badge in the mid-1960s; about 15,000 scouts had qualified for the award by the end of 1969.[18] The AEC also assisted colleges and universities in developing programs in nuclear science and engineering.[19]

The AEC's media campaigns reached millions of people. AEC film libraries loaned films to tens of thousands of groups per year. The AEC's film catalogs listed such items as *The Atom and Eve*, which showed how nuclear electricity would make life easier for all the little Eves of the world by powering toasters, hair dryers, and washing machines, General Electric's film *A is for Atom, Return to Bikini, Go Fission, No Greater Challenge,* and *The Mighty Atom.* In 1960, for example, an estimated 2,500,000 people viewed these films.[20] TV showings greatly increased this number. The AEC's annual report for 1968 said that "AEC films were used widely on domestic and foreign television. U.S. audiences estimated at 23 million viewed AEC films through 276 reported showings on educational and commercial TV channels."[21] Millions of kits of atomic energy information literature were distributed to elementary, high school, and college students. More than 8 million booklets in the AEC *Understanding the Atom* series were distributed between 1962 and 1969.[22] Some editions were translated into several foreign languages, and seven were printed in braille.[23]

Traveling exhibits were an important part of the AEC's PR drive. These exhibits and demonstrations were made "available to qualified exhibitors free of rental and transportation charges."[24] AEC exhibits bore titles such as "Power Unlimited," "Fallout in Perspective," and "The Useful Atom." They included the "five mobile walk-through 'Atoms at Work' exhibit vans," as well as the "large manned 'You and the Atom' exhibits [that] were used during the Summer and Fall to satisfy the needs of State Fairs and larger population centers."[25] In addition, "unmanned package-type 'Atoms in Action' exhibits were booked heavily . . . at schools, conventions, county fairs, and museums."[26]

One of the AEC's more ambitious road shows was called "Your Stake in the Atom." The "prototype unit" of this " 'multiple purpose' exhibit" was completed in the summer of 1961. The AEC reported that

> the new exhibit is housed in an unusual "*Exhibidome*," a 50-foot diameter aluminum geodesic framework which supports a plasticized nylon canopy. Visitors to the Exhibidome find the perimeter lined with instructive animated exhibits and periodically are shown an unusual three-screen, multiple-sequence motion picture and a short stage demonstration which increase their understanding of atomic energy. As "Your Stake in the Atom" tours a State, the Exhibidome is sited in major commuities for approximately a week. At the same time, trained demonstrators traveling with the exhibit present assembly programs on atomic energy to the students of neighboring high schools, before civic clubs, and other organizations. The exhibit also includes a special literature display which is set up at a nearby library to encourage reader interest in books and periodicals dealing with atomic energy.[27]

The AEC's "This Atomic World" lecture-demonstrations were presented to millions of high-school students: in 1967, for example, they were shown to 1,654,000 secondary school students and their teachers.[28] AEC demonstrations and exhibits in the U.S. were viewed by more than 4,800,000 people in 1967.[29]

At the World's Fair in New York in 1965, the AEC's exhibits were seen by more than 2,500,000 visitors.[30] The AEC reported that "Atomsville, U.S.A.," a "highly animated and colorful children's exhibit," was "one of the most popular features" of the Fair.[31]

The AEC's exhibits were popular overseas as well, and huge crowds turned out to see them. "Atoms in Action" demonstration centers, which toured foreign countries, were housed in 10,000-square-foot inflatable buildings.[32] Sometimes the visitor turnout was remarkable. In San Salvador in 1965, 95,000 people attended an AEC exhibit, some 33% of the city's population.[33]

In the 1960s, when utilities began to order nuclear power plants, the AEC assisted the utilities in easing public concern. In its 1966 annual report, the AEC said that

the AEC's public information staff has responded to the surge in utility interest in building nuclear power plants with assistance to utilities in connection with their programs to *improve public understanding* of nuclear power. This includes providing informational materials for informing customers, employees, and for use of local news media to explain the operational and safety aspects of nuclear plants in non-technical language. Members of the staff have participated in regional workshops for utility executives on public informational aspects of nuclear power and have held meetings with utility executives on the same subject.[34]

The commercial nuclear industry established its own public-relations apparatus as well. This PR machine today includes the public-relations departments of the reactor manufacturers (Westinghouse, General Electric, Combustion Engineering, and Babcock and Wilcox), the trade associations (such as the Edison Electric Institute and the Atomic Industrial Forum), and the individual electric utilities.

One important facet of the nuclear PR strategy is developing favorable attitudes about nuclear power in the communities where reactors are proposed and built. Establishing the right attitude within the local communities is important preventative medicine that can save utilities public-relations headaches. Metropolitan Edison's effort in the early 1970s to show that their new station at Three Mile Island was a good neighbor is typical of these community-relations campaigns.

Well before construction was completed on Three Mile Island, Metropolitan Edison set up the "Three Mile Island Observation Center," located in a "beautifully landscaped" park, ideal for picnicking, fishing, and boating. Met Ed's advertisements invited visitors to tour the center, where they could "see displays and hear explanations that provide a basic understanding of the nuclear generation of electric power." Visitors were also "told how nuclear power came to Three Mile Island, how a nuclear plant's wastes are handled and how the environment remains undisturbed."[35] By 1972, more than 75,000 people had visited the center.[36] As one Met Ed brochure put it, "A nuclear generating station like Three Mile Island, along with its Observation Center and Park, promotes tourism, stimulates scientific education and fosters recreational development."[37]

Metropolitan Edison also put together science conferences for high-school students and teachers. These conferences, which were attended by more than 1,000 students and 120 teachers, covered such topics as nuclear medicine, "the energy dilemma," and the handling and transport of radioactive materials, offering "students and teachers a chance to learn more about the rapidly changing world of science."[38] The company also sponsored an "Atomic Energy Merit Badge Clinic" for the Boy Scouts.[39] Met Ed advertised the Observation Center, the science conferences, and the merit badge clinic on local radio, thus heightening their public-relations impact by reaching many people who did not attend any of the events or visit the center with the soothing message that the Three Mile Island plant was a good neighbor.

A second part of the nuclear PR package is the effort to maintain good relations with the media. Hal Stroube, of Pacific Gas and Electric, one of the early nuclear PR strategists, was emphatic about the potential of an *"enlightened press"* to become the industry's "greatest weapon" in gaining public understanding of the atom. Speaking at an American Nuclear Society conference in 1965, Stroube outlined his strategy for gaining the confidence of "the gentlemen of the press":

> No *one* person, no *one* thing, is more important to us in the area of public understanding. We may have speakers' bureaus and films and brochures galore, but by and large the greatest amount of our direct contact with the public is through the news media—newspapers, television and radio. . . . I can tell you with assurance that much of our success in getting our story to our customers and slowly but surely winning their confidence resulted from *first* winning the confidence of the press. We spent countless hours with the editors, science writers and newsmen on dailies and weeklies around our 94,000-square-mile system, and with the newsmen on radio and TV. We answered their questions *honestly* and *fully*, explaining and backgrounding wherever and whenever possible. After all, a newspaperman has to be an expert in many things. . . . Whatever he's covering, he wants *facts*. Given those, he knows how to convey them to his reader or listener or viewer. He got them from us, and when he found he could depend on us for facts and not a lot of hokum, he came to us with increasing frequency to get our side of the controversial story. We won the respect of the news people on our system, and with it we won entree into the minds of our public through the news media. Even more impressive and encouraging, we found that newsmen paid less and less attention to the "anti's" as they found that *their* stories and accusations and suspicions just didn't jibe with the facts which *we* gave them [all emphases original].[40]

Stroube developed strategies for eliminating facts, visual images, and language that stood in the way of *public understanding*. Speaking at the same conference, he suggested that the AEC cancel a study on the hazards of reactor accidents that was then under way.[41] He was afraid that the study would provide ammunition for anti-nuclear activists. The AEC did suppress the study, and it was not published until 1973, when a Freedom of Information Act suit forced its release (see chapter 9).

Stroube also recommended that "some of the *visually* [original emphasis] frightening components of atomic information be eliminated by the AEC. Surely we need no longer glorify the mystique of the atom by photos and films of white-gowned men in surgical caps and influenza masks moving nuclear fuel across the floor of a refueling building for insertion in a nuclear reactor."[42]

Stroube also discussed at length the need to eliminate "*objectionable words* from the atomic lexicon," suggesting that the AEC come up with some "*palatable synonyms*" for "*scare words*":

> Tuesday afternoon some discussion by AEC officials at this meeting touched on the need for replacing the word "*hazards*" in the title of the so-called "Preliminary Hazards Summary Report." I suggest that these gentlemen have the *responsibility* as well as the *ability* to make that change *within a week of returning to Washington.* And while they are at it, I would add that the time has come for some semantic soul-searching about the need for elimination of some other objectionable words from the atomic lexicon. Words such as "criticality"—"*poison* curtain"—"nuclear excursion"—"scram"—"maximum credible accident"—spring immediately to my mind, and I could list a dozen others if given time. My suggestion isn't original. Many others have expressed this same need, but no one has done anything about it. *Some*one has to start *some*where, and I would suggest that the best starting place is in the official language of the AEC. Surely the agency's Public Information Section has the talent and manpower to make a study of this suggestion and hopefully recommend some palatable synonyms for the "scare words" which make our job of public understanding more difficult [all emphases original].[43]

At the time when Stroube outlined these tactics for manipulating the image of the atom, public support for nuclear power was high in most areas of the country. But in the 1970s, the industry came under increasing political attack; in response, it increased its public-relations efforts.

Behind any modern public-relations campaign is a mass of data from public-opinion surveys. PR specialists use these data to determine what people's opinions are, why they hold them, and how people can be manipulated. Points of strong support and weak spots are identified, and the strategists figure out what to say to whom. We rarely get a chance to see the strategy that underlies a new advertising or PR campaign, since most survey data are kept confidential. In 1977, however, a memorandum outlining the strategy for a "*nuclear acceptance campaign*" was leaked to the press. The memo had been prepared in December 1975 for "The Electric Power Industry" by Cambridge Reports, Inc. Cambridge Reports is one of the nation's leading polling firms; its former part-owner, Patrick Caddell, served as

President Carter's pollster. The Cambridge Reports memo provides a unique look at the kind of thinking that PR specialists do.

The Cambridge Reports memo was based on information from national surveys of public attitudes conducted throughout 1975. It outlined a "short-term strategy for stopping the erosion we've seen in the last year in support for nuclear power."[44] The "absolutely critical point to keep in mind" when constructing a public-information campaign, the memo said, is to *use the right medium to communicate the right message to the right target audience* [original emphasis]."[45]

The memo proceeded to apply this *audience/message/medium* formula to the problem of *"stabilizing" people's attitudes* about nuclear power. The company's research showed that support for nuclear power in 1975 was lowest among women, the less educated, lower-income people, the young, and blacks. Lumping these groups together into "three basic categories— women, young people, and low socioeconomic status people (low SES groups)"[46]—the memo turned to the question of messages. A different *message* was developed for each *"target group"*:

☐ "Women must be convinced of the basic safety of nuclear power. That they and their children are not in jeopardy as a result of this technology."

☐ The memo suggested that the industry try to recast the nuclear debate into a debate about the desirability of continued economic growth. People should be told, Cambridge Reports argued, that increased energy use was necessary to power economic growth and that nuclear power was necessary to supply that energy. This was particularly true of the young, who "have to understand better the whole *anti-growth/pro-growth* dialogue. It is our feeling that many of the young who support anti-growth do so with little awareness of its consequences for themselves and others. . . . Messages directed to the young must help bring about understanding in this area and make them aware of the costs and trade-offs involved in *no growth*."

☐ "Low SES groups . . . can be reached most easily. To a large extent, their problem is one of just lack of information. In a world of limited knowledge, they are particularly sensitive to *scare stories* and arguments by the opponents of nuclear. . . . Particular attention here must be given to the *energy supply/growth/jobs linkage*. It is these people who will pay the price of the *'new Puritanism'* of no growth and this must be communicated to them."[47]

On the question of *medium*, the memo advised that "the best spokesmen for nuclear energy are *scientists* [original emphasis]. . . . The public has faith in science, believes scientists and would listen. . . ." The campaign should "put articulate scientists out front, and all spokesmen should emphasize whatever scientific credentials they have."[48] In addition to this piece of general advice, Cambridge Reports suggested a specific approach for each target group:

☐ "For women: Here use other women as spokespersons. Keep in mind the data we've collected showing that proximity to nuclear power plants lessens fear. From this it follows that you should use as pro-nuclear speakers women who live near facilities and can talk, from personal experience, of their safety."

☐ "For young: Again, for this group, it makes sense to stay within the peer group with some of your spokesmen, i.e., young, *ecology-minded scientists* who talk directly, straight, but bluntly about the trade-offs involved in no-growth."

☐ "For blacks: Special advantage here can be obtained by first getting the national black leadership (political, educational, labor, church) to become more aware of the problem. Once this is done, encourage and help them to talk to their communities. In short, we're suggesting here a strategy of working from the leadership on down."

☐ "For low SES groups: These groups too can be reached in part through a leadership-on-down strategy, particularly labor leaders. But, in addition, it is essential for this group that the spokesmen and the message be *blunt, factual and hard-hitting.* Plus to the extent one uses high-cost television messages, place them in time slots heavy with low SES audience."[49]

The memo proceeded to suggest that the campaign should try to shift the nuclear debate onto different ground:

The nuclear debate should *not* be "Should we build nuclear power plants?" but rather, "How do we get the energy/electricity needed for *your* jobs and home?" [original emphasis].[50]

It closed by stressing that the industry needed to go on the offensive, and that attitudinal data should continue to be collected to monitor and refine the PR effort: "Only with on-going research can one measure his effectiveness and act accordingly."[51]

The PR strategy which the nuclear industry has been following since 1975 is in line with the recommendations of Cambridge Reports. The Atomic Industrial Forum, the nuclear industry's trade association, expanded its fledgling task-force *Nuclear Energy Women (NEW)* into a national network in 1976. NEW takes the industry's message to women's groups, using public speakers, publications, conferences, and exhibits. Utility companies have run many ads that stress the theme that nuclear energy is necessary for jobs. One magazine ad, for example, featured a picture of an unemployment line with the heading "NUCLEAR ENERGY WOULD LIKE TO PUT ONE NEW ENGLAND BUSINESS OUT OF BUSINESS."[52]

Westinghouse, the nation's leading manufacturer of reactors, expanded its public-relations operations in exactly the direction that Cambridge Reports proposed. As Gordon C. Hurlbert, president of Westinghouse's Power Systems Company, wrote in *Public Utilities Fortnightly* in 1979:

> Using . . . well-founded [public survey] information, we at Westinghouse have developed effective programs such as our *CAMPUS AMERICA*, in which young nuclear engineers journey on their own volition to campuses to discuss nuclear energy. They usually also appear on local television at the same time. We have other programs which provide radio and television materials and guests anywhere in the country. We have strong programs which try to fill the very real desire on the part of newspaper editors and reporters for *solid information*. And we have programs which appeal directly to minorities.[53]

In an article entitled "Addressing the Nuclear Controversy on University Campuses," published in 1977 by the International Atomic Energy Agency, two Westinghouse officials described the objectives of the Campus America program: "Frankness and openness in responding to questions are an absolute necessity in communicating positively with the university audiences. The objective is to communicate with peers, not repeat *not* [original emphasis] to be viewed as corporate spokesmen who are merely presenting a corporate message in a novel package."[54]

In the late 1970s, the industry began trying an additional tactic: sponsoring and assisting what it calls pro-nuclear citizens' groups. These groups are designed to counter grass-roots efforts by nuclear critics and to create a *pro-energy climate*. In a *Fact Sheet* published in May 1979, only a few months after the accident at Three Mile Island, the Atomic Industrial Forum touted these "new voices" of support for nuclear power that had *"emerged"* in 1978: "The *'silent majority'* began to be heard in state, local and federal hearings around the country. Nearly 100 citizen pro-energy action groups, particularly in the Northeast and West, organized to counter many of the no-growth, anti-nuclear arguments which until recently seemed to dominate the public debate over energy."[55]

In the summer of 1979, an industry-sponsored PR organization called the *Committee for Energy Awareness* prepared a report entitled "To Those Interested in Supporting Citizen Action." The report discussed strategies for using "pro-energy" citizens' groups, stressing that *"citizen action"* can be very helpful for achieving "corporate goals": "Citizens can provide credible, *non-industry spokespersons* able to reach decision makers, *educate the public* and challenge the opposition more effectively than industry."[56] In return, the report argued, industry should supply citizens' groups with resources and funds:

> As a minimum, the commitment by the company wanting to effectively support pro-energy activities must contain the following:
> □ staff support time to aid the citizen group,
> □ information resources,

☐ mailing and secretarial time,

☐ printing or xeroxing, and

☐ money for direct contributions (telephone, rentals, etc.).

The list can be expanded based on the group's needs and the corporate goals.[57]

The back of the report contained a directory of more than one hundred "pro-energy citizens' groups." Calling such groups *citizens' groups*, when they are receiving significant industry assistance and funding, is stretching the meaning of that term. Most of these groups are very closely tied to the nuclear industry. As Westinghouse's Kenneth D. Kearns wrote in an article called "Citizen Action: A Key in the Nuclear Controversy," which appeared in the Committee for Energy Awareness's report,

The following is a partial list of the types of existing advocate groups.

a. Existing pro-energy grass-roots groups.

b. Employees and wives of vendors and utilities.

c. Employees and wives of nuclear suppliers.

d. Employees and wives of large energy users.

e. Professors of engineering and nuclear engineering at universities.

f. Engineering societies, particularly local chapters (such as local ANS [American Nuclear Society] sections).

g. Local American Nuclear Society student chapters.

h. People in business.

i. Etc.[58]

Corporate sponsorship of pseudo-citizens' groups is not new. Nor are any of the public-relations techniques which we have mentioned. Some of these techniques may appear to be inconsequential or even silly. But there is no doubt that the widespread use over many years of these techniques has been a powerful force in perpetuating the nuclear mindset.

PART THREE

Radiation: Managing the Political Fallout

Upon review of the data, the Subcommittee found that while the government was aware of the health hazards posed to the people living downwind from the test site, the government failed to provide adequate protection for the residents of this area during its operation of the nuclear weapons testing program. At the very least, the government owed these people a responsibility to inform them of the exact time and place of each test and the necessary precautions that should have been taken to protect their health and safety. Absent such notification, and uninformed of the evidence held by the government which suggested that exposure to nuclear fallout was causing harmful effects, the residents of this area merely became guinea pigs in a deadly experiment.

from The Forgotten Guinea Pigs, *a
report by the House Subcommittee
on Oversight and Investigations
1980*[1]

CHAPTER SEVEN

Hotter Than A $2 Pistol

The biggest public-relations problem that the AEC ever faced was fallout. In the early 1950s, the AEC's nuclear weapons tests in the Pacific and in Nevada made fallout both a domestic and an international public-relations problem. As thousands of pages of AEC documents declassified in the late 1970s reveal, the AEC's concern over the public-relations impact of its weapons testing outweighed its concern for protecting the public health from the effects of these tests.

In designing its PR campaign, the AEC was guided by a mirror image of the *need-to-know* principle for managing classified information. Instead of a need-to-know, the AEC had a "need-not-to-know" about the potential health effects of its weapons testing. The best way to protect a secret is to keep it from everyone—even oneself. Coverups are unnecessary if you don't know what you "need-not-to-know." The AEC simply failed to gather important data about fallout and its effects; it ignored information suggesting potential problems and focused attention on the wrong problems instead; it harassed scientists who questioned the safety of the tests; and it suppressed scientific research on the health effects of fallout. Throughout these activities and nonactivities, the AEC used classification to keep the public uninformed and to impede the flow of information within the AEC itself. According to a 1979 congressional investigation of the AEC's weapons tests in Nevada, held by Representative Bob Eckhardt's Subcommittee on Oversight and Investigations, "all evidence suggesting that radiation was having harmful effects, be it on the sheep or the people, was not only disregarded but actually suppressed."[2]

The AEC continued this "need-not-to-know" policy about fallout (and the effects of low-level radiation in general) for the next two decades. Testifying before the Eckhardt hearings, Peter Libassi, general counsel for the Department of Health, Education and Welfare, identified several reasons why the American people "were simply not informed of the uncertainties" that the government was aware of in the 1950s and 1960s as to the effects of fallout and low-level radiation.[3] In 1978, President Carter had appointed

Libassi chairman of a special Interagency Task Force on the Health Effects of Ionizing Radiation. Libassi had also been involved in investigations of the health effects of fallout in Utah, during which HEW released 40,000 pages of unpublished material.[4]

Libassi told the hearings that people were "not informed" because there was:

☐ an "overriding concern about *national security*"

☐ only "a single agency . . . managing the entire weapons testing program, rather than multiple agencies"

☐ "a high degree of secrecy and classification of documents"

☐ only "one congressional committee [the Joint Committee on Atomic Energy] instead of several congressional committees performing oversight"

☐ *"limited knowledge*, but . . . also a reluctance to face the hard questions."

Libassi concluded that the uncertainties had remained hidden for so long because there was

a general atmosphere and attitude that the American people could not be trusted to deal with the uncertainties, and therefore the information was withheld from them. I think there was concern that the American people, given the *facts*, would not make the right *risk-benefit judgments*.[5]

The Eckhardt hearings established that the AEC's mindset about fallout led it to endanger the lives of thousands of people in Utah and Nevada. A less cavalier attitude toward fallout might also have restrained the Department of Defense from some of its more ill-advised military experiments with American servicemen. Several hundred thousand troops and civilian personnel were exposed to fallout during weapons tests in Nevada and in the Pacific,[6] including exercises in Nevada where *pentomic units*—nuclear combat troops—conducted maneuvers near detonation points only minutes after explosions.[7]

By 1979, the results of the AEC's policy on fallout were clear. The Eckhardt subcommittee concluded that the government had "totally failed" to protect the residents around the test site, even though there was "sufficient information available" before the tests started to indicate that "people living nearby needed protection."[8] By 1979, medical evidence was mounting that showed that residents around the Nevada Test Site, especially those in southern Utah, were contracting cancer at higher-than-normal rates.[9] There was similar evidence of elevated cancer rates among soldiers who were intentionally exposed to fallout during *atomic battlefield* exercises.[10]

On the domestic front, the AEC's public-relations problem had really begun in 1950, when the National Security Council ordered the AEC to find a continental nuclear weapons test site. Until 1951, the U.S. conducted all of its weapons tests, with the exception of Trinity, in the Pacific, far enough from the mainland to forestall public concern about the resulting fallout. Impetus for a domestic site grew after the first Russian atomic-bomb test in 1949 and the beginning of the Korean *police action* in 1950. The National Security Council wanted a site that was a more convenient commute for the weapons designers at Los Alamos, as well as one more militarily secure than the Pacific islands were thought to be. The AEC knew that weapons tests within the U.S. would be much more dangerous to U.S. citizens than its Pacific operations. The commission doubted whether the American people would share its perceptions of the dire need for a mainland test site.

President Truman's 1950 decision to develop the hydrogen bomb led to international public-relations problems for the AEC as well. Testing of hydrogen weapons, which were potentially 1,000 times more powerful than the Hiroshima atomic bomb, might carry much larger quantities of radioactive materials into the upper atmosphere than the testing of atomic bombs had, and high-altitude winds might then spread fallout around the globe. Unless other countries were reassured that fallout was not hazardous, it was unlikely that these countries would appreciate being involuntarily exposed for the benefit of the U.S. weapons program.

The AEC quickly embarked on a program to guarantee public acceptance of a domestic test site. In a secret December, 1950 memorandum ("classification canceled with deletions," November 29, 1978) recommending the selection of the 3.5-million-acre Las Vegas-Tonopah Bombing and Gunnery Range in Nevada as the new test site (the Nevada Test Site, or NTS), the AEC emphasized the importance of information management to ensure public acceptance of the site. The memo said, "Not only must *high safety factors* be established in fact, but the *acceptance* of these safety factors by the general public must be insured by *judicious handling* of the *public information program.*"[11]

The minutes of a secret meeting at Los Alamos ("classification canceled with deletions," January 29, 1979) in August, 1950 show that AEC officials knew they were taking risks with populations living around the test site, and that they did not fully understand the nature or extent of the risks. There had only been eight U.S. atomic-bomb explosions at that time, and information about the behavior of fallout clouds was very sketchy. Enrico Fermi was particularly outspoken at this meeting on the tentative nature of the group's calculations and conclusions. According to the minutes, "Fermi felt that our conclusions should stress the *extreme uncertainty* of the elements we had to go on. . . ."[12]

Having secretly decided on the Nevada site, the AEC set out to convince the public that the tests were *safe*. At the Eckhardt hearings, Utah Governor Scott Matheson said that the purpose of the AEC's "*all-out public relations campaign*" was

to assure those who lived close to the test site that "there is *no danger*."

That quote was used widely in radio announcements, particularly in the southern communities of Utah, on each day following the actual bomb blasts. Although, on occasion, residents were advised to remain indoors for an hour or two, the AEC announcements were quick to remind the listeners that "there is no danger."

Utahans readily accepted the idea that atomic testing was essential for *national security* purposes and that they would suffer *no harmful effects* because of the testing. They believed the Atomic Energy Commission and its extensive public relations campaign.[13]

One of the most insidious elements of the AEC's public-relations campaign was the use of two teams of Public Health Service personnel as radiation monitors. One team was public, while the other was secret. The public monitoring teams worked to reassure the residents of towns near the test site. They actually lived in the towns for a few weeks before a test. They were under orders "to become a part of the community," and they pitched in as Sunday school teachers and helpful handymen.[14] A report on one test series praised these activities, noting that "such intimate association with the people in the area was *good practical public relations*, and while it may not have altered completely basic public opinion regarding the test, it at least made the explanations of *zone personnel* [the monitors] more acceptable."[15] These public teams talked at every possible opportunity about the safety of the tests, and showed AEC films on the testing program such as *Atomic Tests in Nevada* and *Target Nevada*, as well as films about the peaceful atom such as *Atoms in Agriculture* and *A is For Atom*. The AEC estimated that everyone in the region had been to at least one talk and seen at least one film.[16]

The secret monitoring teams operated outside of the towns, going into the *hot spots* in the desert where there were higher concentrations of fallout. The public teams almost never wore their protective clothing. But the secret teams, who were seen only by a few sheepherders out on the range, wore full protective gear, with their cuffs and sleeves taped shut to keep out the fallout dust. According to a recent report in a local newspaper, "Sheepmen and others who blundered into the teams with the protective clothing described them as '*men from Mars*,' and reported that their attitude was far more brusque than that of the more casually dressed *rad-safe men*."[17]

The integrity of the Public Health Service was further compromised by what HEW general counsel Peter Libassi described to the Eckhardt subcommittee as "certain rules and understandings that were governing the situation during the 1950s and 1960s."[18] The Public Health Service could not issue public statements about radiation issues, and all press releases had to be cleared with the AEC and the White House. Thus, what the public thought were releases from the Public Health Service were actually disguised releases from the Atomic Energy Commission. According to Libassi, the Public Health Service had information on radiation and health that should have been released but never reached the public because it "was classified under security regulations, and disclosure would have been a criminal offense."[19]

The AEC worked to eliminate unfavorable press comments on the tests. It had little patience with newspaper editors who had the temerity to question its unceasing assurances. One AEC report said that while public relations in most communities were "generally good," there were "some specific areas of difficulty. An example of this is the attitude of the newspaper editor in Tonopah who, contrary to editorial opinion in general, has maintained a *highly critical attitude* toward test activities."[20] In a 1957 interview with journalist Paul Jacobs, this recalcitrant editor, Robert A. Crandall of the weekly Tonopah *Times-Bonanza*, described how the AEC responded to his questions:

> Every time we've had an adverse comment in the paper, . . . or what might be interpreted as adverse by the AEC, I've had a couple of boys—or three or four—come in to see me. They come into the office and their tactic has always been along these lines—"Well, you don't believe that the AEC for a moment thinks that there is *any possible harm* in the tests or that any civilian could possibly be injured in any way?" Then they go on to talk about all the precautions they take.[21]

Crandall said that the AEC sometimes tried to intimidate him with absurd stories, or to "Red-bait" him:

> I recall one time when one of the AEC men said, "Suppose there was some woman living around here who had a weak heart and you were to run a story to the effect that this radiation fallout was harmful. And suppose that she had a daughter or small child that was out there. Do you realize that this woman might suffer a heart attack because of the fact that you were spreading *alarming stories*?"
> At other times . . . they said something like this, "Well, of course the *Communists* would like us to stop the tests, too."[22]

One of the AEC's favorite weapons in its public-relations campaign in Nevada was booklets on the testing program, filled with distorted information about the health effects of the tests. A September, 1951 AEC booklet entitled *27 Questions and Answers About Radiation and Radiation Protection* offered the following answer to the question, "Are Weapons Tests In the United States A Danger To The Public?":

> No. The weapons tests, of course, release a great deal of nuclear radiation, but they are carried out in *extremely isolated places* at *safe distances* from human habitations. *No danger* exists from the explosion itself, so far as the public is concerned. The United States has *carefully studied* the results of previous tests and of the explosions at Hiroshima and Nagasaki in Japan. These studies have shown that radioactive materials from atomic explosions can be a hazard only under certain conditions, and *careful safeguards*

FIGURE 8 Illustration from the "Green Book," a 59-page AEC booklet handed out to residents of Nevada and Utah in 1957. Twenty-four nuclear explosions took place at the Nevada Test Site in 1957. The booklet concluded that the AEC's fallout measurements "have confirmed that Nevada test fallout has not caused illness or injured the health of anyone living near the test site."
Source: U.S. Atomic Energy Commission, *Atomic Tests in Nevada, 1957*, pp. 12, 15.

have been established to assure *adequate safety* under conditions prevailing at test sites. *All necessary precautions* are taken, as they are throughout the industry, to guard against even most unlikely hazards.[23]

In 1957, the AEC distributed thousands of copies of the "Green Book" (it had a green cover), entitled *Atomic Tests in Nevada*, to people around the test site. This book was filled with deceptive, misleading statements about every aspect of fallout. According to the AEC, "Your best action is not be worried [sic] about fallout. . . . Please bear in mind that it is *extremely unlikely* that there will be fallout on any occupied community greater than past *low levels*."[24] By this time, worries about the effects of radiation had led many people to acquire their own Geiger counters to check on the AEC's announced fallout measurements. The Green Book cautioned readers not to worry if readings on such instruments appeared high: "We can expect many reports that 'Geiger counters were going crazy here today.' Reports like this may *worry people unnecessarily*. Don't let them bother you. . . . A Geiger counter can go completely off-scale in fallout which is *far from hazardous*, although the fallout might make prospecting [for uranium] difficult for a few days."[25] The Green Book claimed that the "governing policy" used by test managers for establishing "operating controls" for the tests was "that no shot will be permitted at the Nevada Test Site if it presents an *unacceptable hazard* through fallout on nearby communities."[26]

In 1957, the AEC also began promoting *atom-bomb watching* for vacationers in Nevada. The Sunday *New York Times* travel section featured a story by correspondent Gladwin Hill on how to participate in "the non-ancient but none the less honorable pastime of atom-bomb watching." For the first time, Hill effused, the AEC's test program would "extend through the summer tourist season." The AEC had released a partial schedule of the *shots*, so that tourists could make their plans accordingly. Up-to-date shot times were available in Las Vegas a day or two before each shot, from hotels, the chamber of commerce, and the state highway patrol. Hill provided highway guidance to several vantage points around the test site.

The two principal dangers to avoid, according to Hill, were looking at the explosions through field glasses, and automobile accidents (". . . in the excitement of the moment people get careless in their driving").

As for fallout, Hill said not to worry: "In the *dawn's early light* in the wake of a detonation, the atomic cloud can be seen attenuating across the sky. It may come over an observer's head. There is virtually *no danger* from radioactive fall-out."[27]

Despite the repeated assurances that there was "*no danger,*" the minutes of the meetings of the AEC held in the 1950s show that the commissioners knew that the weapons tests were contaminating towns near the test site. The minutes also reveal the commissioners' callous attitude toward the fallout issue in general. For example, at a February, 1955 meeting, AEC Chairman Lewis Strauss reported that the two Las Vegas newspapers had "rather laughed . . . out of court" a state legislator who had filed a bill asking the AEC to move the bomb tests out of the States. Commissioner Willard Libby, the commission's fallout specialist, then chimed in:

> That is a *sensible view*. People have got to *learn to live with* the *facts of life*, and part of the facts of life are fallout.
>
> Chairman Strauss: It is certainly all right they say if you don't live next door to it [sic].
>
> Mr. Nichols [the general manager]: Or live under it.

Commissioner Murray then observed that "We must not let anything interfere with this series of tests—nothing." Later in the same meeting, Strauss referred to the towns of Pioche and St. George, Utah, both of which the AEC's bomb tests "apparently always *plaster*."[28]

The heart of the AEC's need-not-to-know program was its meager fallout monitoring. According to testimony at the Eckhardt hearings, "actual measurements of doses and dose rates were only made at a limited number of locations, mostly along highways, by a limited number of people."[29] The

AEC justified this approach with the convenient assumption that fallout was *uniformly distributed* downwind. A few monitoring stations were said to be sufficient. But as former AEC fallout analyst Harold Knapp told the Eckhardt hearings, "nature is not uniform. Fallout doesn't go down uniformly.... There are many subtleties to the problem of the unevenness in diets, unevenness in deposition, and unevenness in susceptibility."[30]

In Nevada and Utah, residents reported numerous cases of physical illnesses resembling mild radiation sickness, indicating that *hot spots* might have occurred in populated areas. The AEC rarely measured the residual radioactivity in the areas where people had been stricken.[31] Such measurements would have made it possible to calculate the dose rate when the fallout was fresh. Nor did the AEC collect other information about radiation exposures around the test site. According to Knapp, when he investigated fallout for the AEC in the early 1960s, there "had been no systematic collection of data . . . of the levels in milk and fresh vegetables of various internal emitters such as strontium-89, barium-140, and iodine-131."[32] Before 1955, the AEC did not regularly document the exposure of people living around the site by using personal dosimeters such as film badges.[33] The absence of such data has made it much more difficult to do statistically sound epidemiological studies on the relationship between the population exposures and the subsequent appearance of various diseases, such as leukemia, thyroid abnormalities, and solid cancers. Scientists have been forced to measure radiation levels in bricks made during the early 1950s and in preserved specimens of lizards and mammals collected at the time in an attempt to reconstruct the missing exposure data from the *"bank"* of radiation in the bricks and skins.[34]

In the late spring of 1953, Utah sheepherders charged that fallout from the spring series of weapons tests had caused abnormally high losses of ewes and lambs. The AEC had a monopoly on the information needed to assess this charge: it had all of the field data on reported levels of fallout, and it had the only experimental data on the effects on sheep of eating radioactive materials (radioiodine). Admitting that fallout had killed the sheep would have jeopardized continued use of the Nevada Test Site, and the AEC set out from the beginning to deny responsibility for the deaths. According to the summary report of the Eckhardt hearings, the federal government made a "concerted effort to disregard and to discount all evidence of a causal relationship between exposure of the sheep to radioactive fallout and their deaths."[35]

One AEC employee on the scene in 1953 told a sheepherder that "the easiest thing we could do would be to pay for these sheep, but if we paid for them, every woman that got pregnant and every woman that didn't would sue us."[36]

According to Stephen Brower, who was the county agricultural agent when the first team of investigators into the sheep deaths arrived from Los

Alamos, he was told by the AEC's Paul Pearson that "the AEC could under no circumstance allow the precedent to be established in court that the AEC was liable for radiation damages, either to animals or humans."[37] Pearson was chief of the Biology Branch of the AEC's Division of Biology and Medicine. Pearson also told Brower that the AEC would not pay for the sheep losses, but that it might provide some money —a subtle bribe—to the sheepherders in the form of a "desert range nutrition research project."[38] The research project would be specifically prohibited from looking at radiation issues.[39] Brower said that Pearson appeared to be "torn between his loyalty to his native state and the requirements of his job which obligated him to carry out the official policy."[40]

The sheepherders had every reason to suspect that fallout was responsible for the mysterious deaths of their animals. In addition to the deaths, many sheep had unexplained sores on their faces and around their mouths. Veterinarians and sheepherders had never seen such lesions before.[41] The sheepherders had seen the nuclear explosions and had watched the fallout clouds pass by overhead. One man had been warned soon afterwards by Army personnel that he was "really in a *hot spot*" and told to leave, which was impossible with the slow-moving sheep, who graze as they move.[42] Another sheepherder told the Eckhardt hearings of watching some scientists examine some of his still-living sheep with a Geiger counter: "You get up into the respiratory system, and up in their throat, and it was really, really hot. I heard one of the scientists say to another one, 'This sheep is *hotter than a $2 pistol.*' "[43] (A Public Health Service technician who served on an off-site fallout monitoring team in 1953 said in 1979 that the phrase "as hot as a $2 pistol" was used frequently in 1953 by the PHS monitors who were sent into areas of extra-heavy fallout.)[44]

Several of the first veterinarians to see the sheep when they came off the range reported finding evidence of radiation injury and residual radioactivity. But almost all of the most seriously affected ewes and lambs had died on the trail, before the arrival of the investigators. The AEC classified reports indicating that radiation was responsible for the deaths. Brower, the county agricultural agent, wrote in 1979 that he was with a Los Alamos veterinarian who found high levels of thyroid radioactivity in some sheep and who said that the animals' skin lesions looked like radiation burns produced experimentally on animals at Los Alamos. Brower wrote that this investigator later told him that his report had been classified and that he had been ordered "not to discuss or reveal the contents and to *rewrite* the report and delete statements about the sheep having evidence of radiation sickness and lesions."[45]

Despite AEC pressure, several veterinarians continued to insist that radiation was probably involved, and a November 4, 1953 draft report on the sheep case by the director of the AEC's Division of Biology and Medicine said that there was still "a lack of unanimity of opinion" on the cause of death.[46] But on January 6, 1954, the AEC issued an official report on the case in which all traces of the "lack of unanimity" had been eliminated. There was no mention that some of the veterinarians still believed that fallout was involved. The report said that the external effects seen in the sheep "could be produced singly or in combination by a variety of organisms or agents other than ionizing radiation."[47] The AEC did not identify any "organisms or agents" as being responsible, leaving the cause of death undetermined. The AEC further misled the public by claiming that the Public Health Service, the Bureau of Animal Industry, and the U.S. Department of Agriculture all concurred in its conclusions. This lie stood until the 1979 Eckhardt hearings, when it was revealed that these agencies had not concurred in the AEC's conclusions.[48]

The sheepherders did not accept the AEC's denial and took the government to court in 1956. They proved that the AEC chose *shot times* when the wind was blowing in the direction of the sheep; that the AEC made no effort to locate herds of sheep before shots to warn sheepherders; that the AEC did not use its airplanes to track fallout clouds in case they passed over herds; that the AEC knew sheepherders had been exposed and did nothing about it; and that the AEC did not provide advance warning or information on safety precautions to sheepherders.[49] The government did not deny these allegations, contending that the real issue was not AEC negligence, but whether the amount of fallout was sufficient to kill the sheep. In its defense, the government submitted several experimental studies of the feeding of radioiodine to sheep that appeared to prove its claim.

In reaching his decision against the sheepherders, the judge was strongly influenced by his perceptions of the mindsets of the sheepherders and the AEC officials. Why, he asked, should he "disregard the testimony" of the government's *"experts"*?[50] Dan Bushnell, the sheepherders' attorney, argued that if the AEC officials had found that fallout was the cause, "they were, in essence, condemning themselves. This was an investigation by the department which did the act." Bushnell said that in his opinion, AEC officials "got their conclusion, and proceeded to substantiate it. And they were the men in control."[51]

The judge suggested that perhaps it was the sheepherders, not the AEC officials, who had misperceived the situation. In his decision, he said that it was possible that the sheepherders, having decided after the deaths that fallout was the cause, then "endeavored in good faith to reconstruct the situation for the benefit of the investigators" according to their "preconceived notion." The sheepherders were "laboring under the natural desire to protect their property interests." Forced to choose between the two sides, the judge concluded that "I can't bring myself to feel that the effect of trying to arrive at a preconceived notion and justify it would operate any more strongly, and probably it wouldn't operate as strongly upon trained scientific minds as it would upon sheep owners. . . ."[52] The judge dismissed the case against the government, creating a strong precedent that has survived into the 1980s.

In 1979, Harold Knapp conducted an extensive review of the sheep case for the Eckhardt subcommittee, paying particular attention to the scientific evidence presented to the court. Knapp concluded that the government's case "was prejudiced by critical omissions, distortions, and deceptions concerning experimental data on the effects of ingested radioactivity on sheep which was in the possession of the Atomic Energy Commission at the time it made its investigation into the death of the sheep, and (with the exception of one experiment, the results of which were written up in 1954) prior to the 1953 nuclear test."[53]

For example, one study that was reported to the court in 1956 said that the first lambs born to the ewes in one of the AEC's experiments were "normal in size," but that "in subsequent lambing seasons in this group [of ewes] a significant reduction in birth weight of lambs occurred."[54] Knapp located other papers, not presented to the court, which described this same experiment. He was shocked to discover that the first group of *normal in size* lambs developed symptoms of hypothyroidism within two months of birth, indicating that their thyroids had been damaged before birth by the radioactive iodine fed to their dams. And the subsequent lambs were not merely suffering from "a *significant reduction in birth weight*," they were either all born dead or died within a few days of birth![55]

Knapp pointed out that the AEC never asked the critical question in the sheep case: how much fallout would have to be deposited on the desert vegetation so that pregnant ewes would eat enough radioiodine to cause 24% of their fetal lambs to be stillborn or to die within a few days of birth? Knapp suggested that it would have been easy for the AEC to have gathered the data needed to answer this question during its 1955 or 1957 tests by tethering some representative pregnant ewes on typical range, and measuring the fallout levels, the radioiodine levels in the sheep thyroids, and the resulting effects of ingesting the fallout on the ewes and their lambs.[56]

After Knapp's research became public in 1979, the sheepherders moved to have the court reexamine the case. In June, 1981, the same judge who accepted the AEC's position in 1956 announced that he would allow attorneys for the sheepherders to begin an inquiry into whether the government had committed "fraud on the court" in its presentation of the evidence in its possession in 1956. The ruling could open the way for a retrial of the 1956 case.[57]

The AEC was no more willing to take responsibility for causing disease in human beings than in sheep. During the 1950s and 1960s, the AEC systematically interfered with medical and scientific studies of the effects of its weapons testing on the people of Utah and Nevada. According to HEW general counsel Peter Libassi, the AEC withheld information from the public on a number of studies of leukemia and thyroid abnormalities in people who were around the test site. In going through the HEW and PHS files, Libassi said that he had found "specific references to meetings which were held involving the Public Health Service and HEW and White House officials in which reports were discussed and then not released." One HEW secretary wrote a memo to the White House urging full disclosure and full announcement of the issues and questions that had come up about low-level radiation. According to Libassi this memo was ignored: "The press releases which were finally issued did not contain the information and warnings and concerns and questions that were already emerging."[58]

One of the more striking examples of the AEC's attitude toward fallout protection was its handling of radiation exposure from internal emitters. The radium dial case in the 1920s (see chapter 1) had shown that internal exposure from minute quantities of radioactive materials—internal emitters—could be deadly. However, the AEC based its fallout exposure standards on measurements of external exposure—radiation hitting the body from sources outside it. The AEC assumed that if the measured external exposure was below its established limit, then there was no danger of exposure from internal emitters. When one of its own analysts established that there could be high levels of internal exposure despite external measurements within the standard, the AEC tried vigorously to suppress the research.

Radioiodine (I-131) is usually the most dangerous element in fresh fallout because the human body concentrates radioiodine selectively in the thyroid.[59] The food chain from air to grass to cows to milk to humans is an important pathway for delivering radioiodine deposited on the ground as fallout to the human thyroid. On the ground, the radioiodine delivers only a very small external dose to human beings nearby, but that dose can be greatly magnified to the thyroid, if the people in the area drink fresh milk. By 1953, the AEC was aware that milk produced near the Nevada Test Site might be a hazardous substance, although the AEC focused on the *bone-seeker* strontium-90, which behaves in the body like radium, and not on iodine-131.

The AEC did collect some milk, however. After a heavy fallout on St. George, Utah on May 23, 1953, Frank Butrico, the off-site radiation monitor for St. George, wanted to test milk from the area. In his official report, Butrico said he "was afraid it might create a disturbance, should it become generally known that we were collecting milk samples for analysis." The people of St. George were upset about the particularly heavy fallout, and Butrico felt "it would not take much to start *wild rumors*." He finally bought some milk in a local store, but did not record his analysis of the milk. Butrico concluded his report by calling for *"educating the people"* around the NTS: "Most of them are not aware of the *precautions* being taken to *safeguard* them."[60]

In 1963, Harold Knapp, working in the AEC's fallout analysis branch, showed that the failure to consider radioiodine as an internal emitter had resulted in gross underestimates of exposure, especially exposure of babies and children. Knapp did some calculations to make up for the lack of any actual measurement of radioiodine in milk. Taking the AEC's published estimates of the external radiation dose, he used estimates of the amount of radioiodine in fresh fallout, the uptake of radioiodine from pasture by cows, the passage of that radioiodine in cows' milk, the amount of milk consumed by adults and infants, and the behavior of iodine in humans to arrive at an estimated dose to human thyroids. As he told the Eckhardt hearings, this analysis indicated that the AEC "had missed by a factor of a 100 to a 1,000, perhaps, the doses to the thyroid of infants and young children who drank milk from cows that were grazing downwind in the fallout areas around the Nevada test site."[61] It was possible that people had received very high thyroid exposures even though the external radiation levels were within the AEC's guidelines.

Knapp's superiors were very unhappy with his work and tried their best to suppress it. "The people in the Division of Operational Safety were pretty much aghast," Knapp testified in 1979. "I went to talk to the person primarily in charge of estimating health hazards on the technical merits of what I was doing and was told, 'Harold, you are playing with dynamite.' "[62] Knapp asked Gordon Dunning, the deputy director of the Division of Operational Safety, for some radiation exposure data for certain towns. Knapp wrote that he was told that there were "*no such data*," although it later turned out that Dunning's office did have these data. Knapp eventually got the information he needed from the Utah Department of Health, who told him they had been forced to build monitoring facilities that duplicated those of the AEC's Off-Site Radiological Safety Organization "because data pertaining to the safety of the citizens of Utah was not forthcoming from the AEC."[63]

Knapp's superiors were worried about the impact of his findings on their public-relations campaign. In a memorandum to the director of the Division of Operational Safety on Knapp's work, Gordon Dunning wrote: "We have spent years of hard, patient effort to establish *good and calm relations* with the public around NTS. Such action as the author's has been harmful."[64] That the AEC's failure to consider the hazards of radioiodine might have been "harmful" was not mentioned in Dunning's memo. Instead, he struggled in vain to come up with a method for disavowing Knapp's paper. How could the AEC publish such findings? Dunning wondered. "It is difficult to see how the Commission can 'admit' to *dire deeds* and in the next breath say we are going to keep on doing the operations."[65]

The AEC held up publication of Knapp's work for months. A special review committee was appointed, an unprecedented step. Dunning hoped this committee would agree to suppress Knapp's study, but the committee did not oppose publication. In his memo to the director, Dunning lamented: "One might argue that the committee was not made *fully aware* of the *pitfalls* inherent in the treatment given the field data. But the fact remains, they did not turn thumbs down. Now what do we do?"[66]

The AEC finally agreed to publish Knapp's study, but only after he agreed to remove all the examples of his calculations for exposures in specific towns. Having fought with the bureaucracy long enough, Knapp agreed to this change, knowing that he would be able to publish an uncensored version of his report later on in the journal *Nature*.[67] In an interview in 1980, Knapp explained that throughout the controversy, he had only tried to do his best as a responsible analyst: "I'm an analyst. The answer's the answer, and I just get the best answer I can get. I don't mean to imply that I can't be wrong. But the whole thing is, do you probe, or don't you, do you use your best evidence, and take it wherever it comes out? It's not possible to do scientific work if the answer is known in advance."[68]

While the AEC published Knapp's edited paper, it also decided to ignore Knapp's conclusions and to make no changes in its policies, for fear of the public-relations impact. In a January, 1963 memorandum to the AEC's general manager, Nathan H. Woodruff, director of the Division of Operational Safety, wrote:

> The present guides have, in general, been *adequate* to permit continuance of nuclear weapons testing and at the same time have been *accepted* by the public, principally because of an *extensive public information program*. To change the guides would require a *re-education program* that could raise questions in the public mind as to the validity of the past guides. Lastly, the world situation today is not the *best climate* in which to raise the issue. Therefore, we recommend the continuation of the present criteria.[69]

After the release of Knapp's study, the AEC continued to resist investigations of the health effects of the tests. For example, in 1965, the Public Health Service decided to do a study of children in Utah that included clinical examination of their thyroids, the glands that Knapp's research had suggested were most likely to have been injured by fallout. In an August, 1965 memorandum, Gordon Dunning wrote to Dwight Ink, the acting general manager of the AEC, bemoaning the *"fanfare and publicity in Utah"* that would attend the announcement of the PHS study. Dunning said that "it might not be wise to attempt to stop the PHS studies," but suggested that the AEC technical staff prepare an attack on "the *fallacies* upon which the studies are being based." Such an attack, Dunning hoped, might avert "a *potential fallout scare* by placing the purpose of the studies in *proper context*."[70]

In September, 1965, Ink sent a memorandum on the PHS study to the AEC commissioners that reveals why the AEC was so concerned about the PHS study. Ink concluded that

> although we do not oppose developing further data in these areas, performance of the above U.S. Public Health Service studies will pose *potential problems* to the Commission. The problems are : (a) *Adverse public reaction;* (b) Law suits; and (c) Jeopardizing the programs at the Nevada Test Site.[71]

The Public Health Service, after consulting with the White House, finally began the children's study, but according to Dr. Joseph Lyon, an epidemiologist at the University of Utah and director of the state cancer registry, the PHS study was poorly designed. It did not answer the questions raised by Knapp's study. It did not ask where the children had obtained their milk, making it impossible to estimate the radioiodine doses to their thyroids. In an interview, Lyon compared this omission with "doing a study of lung cancer and forgetting to ask people if they smoked."[72] The study was never completed and thus reached no conclusions about the impact of fallout.

Only after the AEC was officially abolished did congressional investigators and independent scientists begin to unravel the history of placing public relations before public health in the operation of the Nevada Test Site.

BRAVO

Fallout became an international public-relations problem that the AEC was never again able to quiet after the March 1, 1954 BRAVO test. This 15-megaton hydrogen-bomb explosion produced lethal levels of fallout over a 7,000-square-mile area of the Pacific, killing a Japanese crewman on the fishing boat Lucky Dragon and heavily contaminating a group of American personnel and Marshall Islanders.[73]

Before BRAVO, many AEC and military officials believed that hydrogen bombs would not produce much fallout. This belief was supported by the results of the first large thermonuclear test, the MIKE shot, made in late 1952, which had a yield of 10.2 megatons. Some scientists had predicted a large amount of fallout from MIKE. But when they asked to do fallout monitoring more than 50 miles from the island site, they were told that they would have to get outside financial support.[74] As a result, very little monitoring was done, and little fallout was detected. According to Merrill Eisenbud, an AEC fallout specialist, there was "a very, very influential group of people both among the military and civilians, who insisted that there never was any MIKE fallout, that it all went up in the stratosphere, and that probably most of it was still in outer space, and there even were calculations to prove it."[75]

The MIKE experience made the planners of BRAVO so confident that fallout was not a problem that they refused a recommendation that plans be made for evacuating the populations of the islands nearest the test site on the grounds that there would not be any fallout. Scientists were able, however, to get the AEC to put an automatic radiation recording instrument on one of the closer inhabited islands; this instrument saved the lives of a group of American service personnel and of the native Marshallese. The meter on the instrument went off the top end of the scale seven hours after the shot, but no one hurried to see what was happening. According to Eisenbud, "there was not enough interest in the Task Force to authorize sending a plane. . . . there was just a complete breakdown as far as information was concerned, in taking the steps that were necessary in order to evaluate the situation. . . ."[76] After thirty-six hours, a plane was finally dispatched. Finding high radiation levels, the plane landed and evacuated the Americans. Not until fifty-six hours after the shot did an American boat arrive to remove the Marshallese.[77]

The U.S. had not warned the Marshallese about what to do if fallout should come down on their islands. During their fifty-six hours of exposure, they received burns on their skin from fallout they could easily have washed off, and large doses to their thyroids and other internal organs from radioactive particles in their food, which they also could have washed off.

The BRAVO story became a public-relations disaster for the AEC on March 16, 1954, when Japanese newspapers reported that the 23-member crew of the fishing boat *Lucky Dragon* had arrived in port suffering from radiation sickness caused by grey, snowlike particles that fell on the boat after the March 1 test. The captain said the boat had been well outside the *danger zone* established by the United States, and that he had received no warning before or after the test from U.S. authorities.[78] Instead of using a prosaic word like *fallout*, the Japanese press evocatively named the phenomenon *ashes of death*.[79] Relations between the U.S. and Japan deteriorated further when Japanese fishing boats began bringing in radioactive tuna caught far from the test site. The Japanese government set up an emergency contamination standard for the *crying fish*, named for the chatter made by Geiger counters responding to the hot fish.[80] The Japanese regarded the incident as a third atomic bombing, and the resulting uproar forced President Eisenhower and AEC Chairman Lewis Strauss to hold a joint press conference on March 31 to discuss the affair.

Strauss tried to downplay every aspect of the BRAVO explosion. He denied that radiation had caused the skin lesions on the Japanese fishermen. He claimed that the burns were "due to the *chemical activity* of the *converted material* in the coral rather than to radioactivity."[81] Japanese scientists were familiar with radiation burns and were outraged by Strauss's false assertion.

Strauss also said that "the *facts do not confirm*" stories about "the widespread contamination of tuna and other fish. . . ."[82] By the end of 1954 (after five more U.S. Pacific weapons tests) 683 Japanese boats had landed contaminated fish, and 457 tons of tuna had been condemned and destroyed, roughly one tuna out of every 200 caught.[83]

Strauss said that there would be no health problems for the contaminated American servicemen nor for the Marshallese: "Today, a full month after the *event*, the medical staff . . . advised us that they anticipate *no illness*, barring of course disease which might be *hereafter contracted*."[84] This last phrase was a handy catchall for whatever disease might appear once the latency period expired. Strauss and the AEC's doctors knew perfectly well that one month was a meaningless period for evaluating long-term radiation injury.

At the time of Strauss's press conference, the Marshallese had lost their hair and had skin lesions on the scalp, neck, shoulders, feet, limbs, and trunk. Some had wet, weeping, and ulcerated lesions, especially on their feet, which had received extra exposure since they went barefoot.[85] These external lesions healed, but the islanders' health problems had just begun. The United States followed the health of the islanders with annual medical surveys. In 1963, the medical survey revealed the first thyroid abnormalities; children were especially hard hit. By 1969, 84% of the children on one island who were under ten years old at the time of exposure had developed thyroid nodules. Some had developed thyroid cancer, and many had to have their diseased thyroids removed.[86]

In a February, 1955 public report, the AEC tried to put the BRAVO test in the best possible light:

> If we had not conducted the fullscale thermonuclear tests mentioned above, we would have been in ignorance of the extent of the effects of radioactive fall-out and therefore we would have been much more vulnerable to the dangers from fall-out in the event an enemy should resort to *radiological warfare* against us.[87]

Such self-serving statements could not quell the international outcry against atmospheric testing. Not even the most vigorous campaign of public relations and information suppression could hide the growing evidence of worldwide public health hazards from continued testing. In 1963, the superpowers yielded to public pressure and agreed to halt their atmospheric testing of nuclear weapons.

CHAPTER EIGHT

Undue Anxieties

The controversy over fallout threatened the plans of nuclear developers around the world to proceed with building thousands of nuclear reactors over the following few decades. A single 1,000-megawatt pressurized-water reactor may contain 15 billion curies of fission products, the equivalent of the radiation produced by the radioactive decay of the 4 billion tons of naturally occurring uranium contained in all the oceans of the world.[1] The accidental release of even a small portion of the fission products of one reactor would be an unprecedented peacetime disaster. The public's concern about fallout spilled over to concern about reactor accidents and about the *planned releases* of radioactive materials throughout the nuclear power fuel cycle.

Nuclear developers responded to these concerns about the public health impact of nuclear power by treating them as public-relations problems, just as they had done with fallout. Members of the public who raised questions about the hazards of radiation were said to be suffering from "emotional" or "pathological" problems.[2] And since the AEC and its successors have controlled most of the funding for radiation research, scientists who did not cooperate with the public-relations program risked losing funding and jobs.

The AEC began consigning its critics to the psychiatric couch early on. In 1948, AEC Commissioner Sumner T. Pike appealed to psychiatrists at the annual meeting of the American Psychiatric Association "to know at least enough about the subject to cool off anyone who seems *hysterical* about atomic energy. . . . It might be well for the security of the country."[3]

In early 1957, the World Federation for Mental Health called for a study of the mental health implications of the *peaceful* use of atomic energy. The federation said that nuclear power might arouse "many *irrational fears*, or irrational degrees of fear . . . because of the very special types of threats inherent in our popular concepts of atomic energy." The public needed "protection from *undue anxieties and fears*." The federation warned that "the techniques of the usual highly skilled public relations divisions will not be adequate" to deal with these new problems. What was needed were "the technical skills of specially experienced personnel . . . who are trained in the interpretation of deep-seated types of reaction in individuals, groups and populations." These people could use their knowledge of "personality dynamics" to handle *"irrational attitudes"* and to build *"positive morale"* about nuclear power.[4]

The World Health Organization responded to the federation's plea in the fall of 1957, convening the Study Group on Mental Health Aspects of the Peaceful Uses of Atomic Energy. The group included both psychiatric physicians and professors, and representatives from the AEC and the European nuclear industry. In describing its mission, the study group concluded that "the behavioural sciences can make a valuable and concrete contribution to the adaptation of mankind to the advent of atomic power, making it indeed as painless and as unharmful as possible and allowing man to reap a rich harvest from the seed his inventive genius has sown."[5] Thus it was no surprise that the study group was alarmed by the public's "irrational fears, irrational hopes, or irrational tendencies to ignore or deny the extraordinary potentialities of atomic power."[6] According to their report, these "irrational fears" were not supported by "present knowledge":

> When the evidence of *abnormal emotional response* to atomic energy is checked against reality it is clear that the response is quite unjustified, in terms both of quantity and quality. The balanced conclusion of a review of present knowledge would be that, even if all the objective evidence were interpreted in the most pessimistic way possible, the weight of evidence would not justify anxiety in the present, and only vaguely and remotely in the future. Yet anxiety exists and persists to a quite extraordinary degree. This can only be accounted for by looking into the psychological nature of man himself.[7]

The study group suggested that there was "a striking analogy . . . between the situation of man in relation to atomic power and that of the very young child first experiencing the world."[8] Clinical experience showed that children were most anxious in situations in which they felt they were at the mercy of what they perceived to be omnipotent powers. Under these circumstances, particularly when "everyday childhood situations . . . as feeding and excretion" were involved, children would often regress to more infantile forms of behavior. The study group claimed that it was therefore no accident that "of all the fears rising from radiation, whether it be from atomic bomb fall-out or from nuclear plant mishap, it is the danger to food which is generally the most disquieting." This same principle also applied to nuclear waste: "As with feeding, so with excretion. Public concern with atomic waste disposal is quite out of proportion to its importance, from which there must be a strong inference that some of the fear of 'fall-out' derives from a symbolic association between atomic waste and body waste."[9]

This explanation is the most ludicrous kind of dime-store Freudianism; it trivializes people's concern about fallout and nuclear war. But the study group was deadly serious about the richness of insight which this crude, narrow-minded analysis provided, as it showed in its analysis of public reaction to a serious accident at a British reactor that occurred while it was meeting. The reactor, Windscale Pile No. 1, began releasing large quantities of radioiodine and other fission products after a combination of human error and poor instrumentation led to its uranium fuel catching on fire. The heat of the fire caused the metal claddings of other uranium-fuel elements to melt, releasing the gaseous fission products. Several hundred square miles were contaminated with so much radioiodine that the British government was finally forced to confiscate milk from the region. In a dramatic gesture, the government dumped the contaminated milk into the sea.

The Windscale accident provided the study group with a perfect opportunity to apply its purported psychological expertise. According to one study group member, the British writer Lord Ritchie Calder, the stories on the first two days of the accident were "*factual and reassuring.*" But on the third day, when the milk was poured into the sea, Calder wrote that "the headlines exploded. . . . As we studied the third day's headlines, Hans Hoff [chairman of the study group and director of the Psychiatric University Hospital and Neurological Institute in Vienna] said to me, 'Obviously all the editors were *breast fed*.' It was, to him, a perfect example of 'regression.' "[10]

In its report, the study group did admit that part of the public's "undue anxieties" might arise from "the fact . . . that the scientific processes involved in the warlike and the peaceful uses of the atom are indistinguishable to the non-scientific mind, and to the scientifically educated person it is clear that many of the basic processes can be used equally for both purposes. . . . It is virtually impossible, for the authorities as well as for the general public, to know where to draw the line."[11]

Given the public's anxieties about radiation, the study group cautioned against arousing new anxieties by overemphasizing the need for safety, both among the public and among nuclear workers. They claimed that experience showed that nuclear power plants could be safely sited in populated areas, and that we would eventually be asking ourselves "whether there may be a critical point in public attitude where a policy of siting plants in remote areas will tend to augment rather than diminish public anxiety."[12] Likewise, in appraising nuclear workers of their job hazards, "there may be a point of elaboration beyond which safety precautions tend to increase, rather than to allay, anxiety."[13]

The study group worried that its advice might be rejected by the public because of "a general *mistrust of information sources*" among the public, a mistrust fostered by "the widespread publicizing of mistrust and disagreements among scientists, not only in connection with nuclear energy, but also in other matters, such as . . . the carcinogenic effects of tobacco."[14] The mental health profession was already suffering from the "known association of psychology, medical and otherwise, with so-called psychological warfare, propaganda and advertising, and suspicions attaching to the profession owing to such techniques as 'brain-washing.' "[15] To improve their credibility, the study group warned scientists to stop vacillating "between statements which are limited to their scientific competence and statements which have the mantle of science but which are actually expressions of value and even of policy decision." In discussing scientific matters, scientists must "avoid any confusion between facts and judgments on facts."[16]

The study group concluded that while the problem of manipulating the anxieties of the pre-atomic-age generations was a difficult one, there was hope for the future: "In the long run the greatest hope of mental health in the future of the peaceful uses of atomic energy is the raising of a new generation which has learnt to live on terms with ignorance and uncertainty and which, in the words of Joseph Addison, the 18th century English poet, 'rides in the Whirl-wind, and directs the Storm.' "[17]

EARLY RETIREMENT

The AEC and its successors dealt harshly with any scientist whose research appeared threatening to the existing radiation exposure standards. One of the best-documented examples of this policy is the case of Dr. Thomas Mancuso, an industrial epidemiologist hired by the AEC in 1965 to study the effects of occupational radiation exposures on workers at Hanford, Oak Ridge, and other AEC installations. In 1974, when Mancuso showed he was unsympathetic to the AEC's public-relations approach to radiation hazards by refusing to issue a misleading and possibly false press release, AEC officials took secret action to terminate Mancuso's funding. These officials then attacked Mancuso's standing as a scientist. A congressional inquiry in 1978 showed that the AEC had lied about the real reasons for Mancuso's firing. The inquiry clearly established that the AEC had acted to achieve political and not scientific ends.

AEC officials never expected Mancuso's work to produce any problems; they believed that it would bolster the commission's claim that worker exposure standards were adequate. For example, in a 1967 memorandum to the director of the AEC's Division of Biology and Medicine, Dr. Leonard A. Sagan of the Medical Research Branch of the division wrote that it was the "unanimous opinion" of a group that had just reviewed Mancuso's study that "aside from a certain *'political' usefulness*, it was very unlikely that new information on radiation effects will accrue from this study."[18]

In the spring of 1974, the AEC became concerned about the possibility that Mancuso's work might be politically damaging, when an independent study by Dr. Samuel Milham, a Washington State scientist, indicated that there was an excess of cancer deaths among men who had worked at Hanford from 1950 to 1971, an excess Milham suggested might be due to their occupational exposure to radiation.[19] Milham had no way to check this hypothesis because he did not have the exposure records for the Hanford workers. Mancuso had been gathering these exposure records.

Milham's study threw the AEC into a panic, and it moved quickly to contain the potential damage. The AEC's first instinct was to have Mancuso publicly refute Milham's finding that there were excess cancer deaths. Mancuso testified in 1978 that

> I was on the telephone by the hour relative to communications from Hanford and from the AEC headquarters, who were very much concerned about the impact on the public should there be any release of Dr. Sam Milham's findings.
>
> They were very concerned about my reaction to what might occur and what should be said or could be said relative to the press, should the press query me on this particular problem.[20]

Mancuso's project officer at AEC headquarters, Dr. Sidney Marks, called up and dictated a press release which he wanted Mancuso to issue. (Marks, we should note, was one of the authors of the Hanford studies of the effects of feeding radioiodine to sheep that were used to cover up the AEC's role in the NTS sheep deaths in 1956.)[21] The release was designed to quell any *undue anxieties and fears* that publication of Milham's study might engender. Marks wanted Mancuso to say that he had seen Milham's paper, but that there was "a *wealth of information*" which he had accumulated during his years of research that contradicted the finding of excessive cancer deaths among the Hanford workers. The release went on to say that the information Mancuso had collected showed that there was "*no evidence*" of radiation-induced cancer or other deaths occurring more often among Hanford workers than among their brothers and sisters who did not work at Hanford.[22]

Mancuso was still gathering data on the Hanford workers and had not finished his analysis of the data. As a responsible scientist, he did not want to report what he felt might be false results from a premature analysis. He ignored Marks's request and did not issue a release.

Mancuso's refusal to compromise his principles as a scientist for the AEC's public-relations program led his superiors to move swiftly to remove him. He had shown he could not be counted on to distort information, and the data base he was working with was large enough that it was possible that he could demonstrate a radiation-cancer link, if one existed, at doses below the maximum permissible limit for workers. Acting in secret, Marks wrote a number of memoranda which resulted in the termination of Mancuso's contract and, more importantly, the transfer of his laboriously assembled data to friendly hands at Oak Ridge and Hanford.

Mancuso's firing created a controversy in Washington: was the firing legitimate, as the government claimed, or was the action politically motivated, as Mancuso and his supporters claimed? A congressional inquiry resulted, during which ERDA and DOE officials lied repeatedly in an effort to hide the real reasons for the firing. (The Mancuso case took place under the successive administrations of the AEC, ERDA, and DOE.) In response to one of the first letters questioning Mancuso's dismissal, Dr. James Liverman, acting assistant secretary for the environment for the DOE, wrote back that the study was being removed from Mancuso's control because of his *"imminent retirement."*[23] DOE spokespeople repeated the "retirement" explanation to the press.[24] But Mancuso had no plans for early retirement. No one from ERDA or DOE ever contacted Mancuso, or the University of Pittsburg, where Mancuso was based, on the question of his retirement. Under the policies of the University of Pittsburg, Mancuso, who was 62 in the summer of 1974, could continue working until age 70 as long as research funds were available to support his work.[25] At the 1978 hearings, Liverman was forced to concede that "as a result of having reviewed the facts in detail . . . my use of the words *'imminent retirement'* was unfortunate, inappropriate, and perhaps, even in error."[26]

Liverman insisted, however, that there were substantive reasons behind his decision to fire Mancuso. He cited a "clear lack of substantive publications appearing in reference journals. Even papers on Dr. Mancuso's methodology for analysis would have been highly useful."[27] But as Mancuso showed the committee, he had published his methodology by special arrangement in a November, 1971 180-page publication of the Health Physics Society of Richland, Washington, since his paper was much longer than most refereed journals normally carried.[28] Liverman also criticized what he said was a "reluctance on Dr. Mancuso's part to initiate any analyses until all data collection was complete. . . ."[29] This criticism was contradicted by the comments of one official review of Mancuso's work which praised his refusal to publish prematurely, stating that since various unexpected biases "are apt to creep into some portions of the data in a study as large as this, patience in checking out results is desirable."[30] A 1977 letter from the manager of the Human Health Studies Program of the DOE's Division of Biomedical and Environmental Research also supported Mancuso's decision not to publish prematurely, noting that because of the long latency periods of many cancers, "it would have been extremely difficult to conduct meaningful analysis at an earlier time."[31]

Liverman's strongest attack on Mancuso's work was his claim that peer reviews of the study had produced a "judgment by his scientific peers that the work should be limited, terminated, or another investigator selected to be the principal investigator."[32] The peer review process, in which a jury of one's scientific peers critically reviews one's research, is a widely used method for evaluating the quality of scientific research and for determining future funding levels. The process is supposed to yield the most unbiased, "objective" analysis possible, as well as providing constructive criticism that can be used to correct deficiencies in the research. As the hearings revealed, the DOE had made extremely biased use of the peer reviews of Mancuso's study.

Liverman lied to the committee about a December 2, 1974 memorandum from his staff which he claimed said that peer reviews of Mancuso's work recommended changing the contractor. Liverman said his decision to terminate the contract was based on this recommendation. But the first few times he mentioned this important document, he was unable to put his hands on it. The reason for his reluctance to read the actual document became obvious when he was confronted by Representative Paul Rogers with its contents:

> Dr. Liverman. I don't have that memo in front of me.
>
> Mr. Rogers. I recall that you said your staff had said in the memo that the peer review had recommended the change in contractor, and that is what you based your decision upon.
>
> Dr. Liverman. I do not have that memo in front of me.
>
> Dr. Marks. Here it is. [Sidney Marks, Mancuso's project officer, who was also on the stand]
>
> Mr. Rogers. Well, this memo is what you told the committee was the basis of your judgment. And I don't see any such referral.
>
> Dr. Liverman. You are absolutely correct on that.
>
> Mr. Rogers. Well, now what is the committee to conclude, Dr. Liverman? You told us that your judgment was based on a staff memo which told you that the peer review didn't recommend that Dr. Mancuso continue with the study. Now we see the memo that you referred us to, and that recommendation simply is not there.[33]

Liverman could only reply lamely that his decision "was based on this memo and on the many other peer reviews, and other factors that are tied in with it."[34] Once again, Liverman was lying. When Representative Rogers pressed Liverman to produce the "many other peer reviews," the only thing Liverman's DOE associates could come up with was a negative recommendation from one of the five reviewers who participated in a 1972 peer review.[35] The committee had already read a letter from another member of this same peer review group which concluded as follows: "It was the consensus of the group that this was a highly desirable research activity, and deserved continued support. We also agree that the University of Pittsburg [i.e. Mancuso] should continue as the contractor."[36] The DOE was unable to produce any peer reviews of Mancuso's work which contained a majority recommendation that his contract be transferred to some other contractor.

The congressional inquiry showed that after deciding to dismiss Mancuso, federal officials ensured that the government would gain complete control over his important data base. Mancuso's data on Hanford workers ended up at Battelle-Pacific Northwest Laboratories, in the hands of Sidney Marks, the AEC project officer who had initially recommended that Mancuso be fired after Mancuso refused to issue Marks's press release.[37] This cozy disposition of the Hanford data also produced a conflict of interest at the institutional level, since Battelle, a major nuclear contractor at Hanford, would be evaluating data on some of its own employees. Mancuso's Oak Ridge data were assigned to the care of Dr. Clarence Lushbaugh at Oak Ridge Associated Universities (ORAU). Lushbaugh was an administrator at ORAU and did not have anyone on staff at ORAU who could handle the project at the time it was reassigned.[38] (Like Marks, Lushbaugh was a veteran of the sheep death case. At Los Alamos, he had been in charge of experimentally burning the skin of sheep with radioactive materials and comparing the lesions produced to those on the range sheep.[39])

In transferring Mancuso's contract to ORAU and Battelle, federal officials followed none of the normal scientific and bureaucratic procedures for such cases. DOE did not issue a "request for proposal"; DOE did not put the contract out for competitive bidding, and no other contractors were given the opportunity to submit proposals; neither ORAU nor Battelle submitted a research design or protocol; and there was no scientific peer review of the new contractors' past work or present capability, despite grave deficiencies at both locations.[40] Battelle, for example, had never done a human epidemiological study before.[41]

After having falsely used the peer review issue to condemn Mancuso, DOE officials insisted that they did not need to peer review their decision to send the most important epidemiological study of radiation to a place that had no experience doing such work. Dr. Walter Weyzen of DOE explained to Representative Paul Rogers why ERDA and DOE had not used peer review:

> Dr. Weyzen. Sir, I feel very comfortable that we can do without it because I know the quality of the people involved in it.
>
> Mr. Rogers. But they have never before studied humans?
>
> Dr. Weyzen. Sir, I interact with people at Hanford, on a weekly basis on technical and scientific matters. I think I can trust my judgment in knowing what the people can do.
>
> Mr. Rogers. Then you are telling me we don't ever need peer review.
>
> Dr. Weyzen. That is not what I am saying. It is your statement, sir.
>
> Mr. Rogers. But it is a pretty logical conclusion, sir, if you say you can make all the judgments that everybody else uses peer review to do.[42]

PART FOUR

One Chance in Five Billion

We screwed up—and I mean by "we" the nuclear community, the vender and the N.R.C.   Davis-Besse was ample warning, and if we had paid ample attention to it Three Mile Island could have been prevented. If you want to use the term "complacency" to describe our behavior, I won't quibble with you, but the term I'd rather use is "mind-set," or "attitude." I'd had the attitude that reactors were fairly forgiving, in the sense that they could withstand a lot of problems without having those problems turn into serious accidents. I don't feel that way anymore.

Denwood Ross, deputy director,
Office of Nuclear Regulatory Research,
Nuclear Regulatory Commission
1981[1]

CHAPTER NINE

Hiding the Hazards

When the AEC began developing nuclear power reactors, it believed that guaranteeing reactor safety would not be an important obstacle to the new technology. At first, distance was the principal safeguard, and test reactors were located well away from cities. But the costs of transmitting electricity pushed the AEC to begin substituting engineered safeguards for distance, and larger and larger reactors were built closer and closer to cities. Some of the most important of these safeguards were not tested before they were set in steel and concrete. Nuclear developers assumed that when and if the tests were ever done, the results would confirm their optimistic mindset about reactor safety. And when doubts arose about this mindset, the government suppressed critical reports, hid the worst nuclear accident in history for sixteen years, intimidated skeptical AEC employees, and prevented the public from learning of questions about safety standards within the AEC itself.

The AEC's mindset about safety was apparent in 1956 in the first contested licensing case, a proposal to build a 100-megawatt *fast breeder* reactor about 30 miles from Detroit. (The reactor would use more energetic— or fast—neutrons than a light-water reactor.) The plant, named after Enrico Fermi, was an experimental design; it would be the world's first commercial breeder reactor. The AEC was anxious to develop breeder technology, since only breeders could *burn the rocks*. But breeders were also more dangerous than light-water reactors: it was theoretically possible that the more highly enriched uranium fuel in a breeder reactor could form a *supercritical assembly* during a *core meltdown*, producing a small nuclear explosion with the force of hundreds of pounds of TNT.[2] Such an explosion could blast open the containment building, releasing a cloud of highly radioactive gases and particles.

Shortly before the Power Reactor Development Corporation (PRDC) applied for a construction permit for the Fermi plant, there had been a *partial core meltdown* at the EBR-1, the AEC's small Experimental Breeder Reactor.[3] The 15-member Advisory Committee on Reactor Safeguards (ACRS), which reviewed the safety provisions of reactors, concluded that "there is insufficient information available at this time to give assurance that the PRDC reactor can be operated at this site without public hazard."[4] Referring obliquely to the EBR-1 meltdown, the ACRS wrote that "the experience that now exists on fast power reactors of high power density is *not wholly reassuring.*" The ACRS felt that the existing data did not prove that *"no credible supercriticality accident* resulting from meltdown could breach the container," and that much more research was needed to produce the experimental support for the PRDC design.[5]

Until after the Fermi case, the AEC conducted its safety evaluations in closed meetings, not releasing reports from the ACRS or the in-house Hazards Evaluation Branch.[6] The AEC kept the ACRS report on the Fermi plant secret and went ahead with plans to issue a construction permit. Regulations allowed the commission to issue a construction permit, as long as it believed that all *unresolved safety problems* would be resolved upon application for an operating permit.[7] When the Joint Committee on Atomic Energy learned of the ACRS report and asked that it be made public, the AEC refused, invoking *executive privilege.* In a letter to the JCAE, AEC Chairman Lewis Strauss maintained that ". . . the independence of the staff of the Commission or advisory committees to the Commission would be very seriously impaired in the future, the value of their contributions would be greatly diminished, and the regulatory functions of the Commission would be correspondingly impeded" if the ACRS report were made public.[8]

The JCAE finally flushed the document out of the AEC, forcing the commission to offer a public explanation of why it had rejected the advice of the ACRS. This explanation, presented in an October 19, 1956 letter from the AEC's General Manager K. E. Fields, clearly defines the mindset which guided the AEC in its deliberations and research on reactor safety.

Fields reported that the AEC staff did not deny any of the allegations about the potential danger of the plant or the lack of experimental data on such plants. But where the ACRS saw uncertainty and urged caution, the AEC counseled *optimism,*[9] an optimism that was based not on experimental data, but on faith that somehow everything would work out right for nuclear development.

Fields admitted that there were still *"unresolved areas of doubt"* and that there was a "considerable difference in the views held by the *experts* on the *degree of certainty"* in predictions about how the PRDC reactor would behave. But Fields said that the AEC believed these problems would be solved because the commission's research program "will go forward."[10] He was thus arguing for trust in research not yet done, which might well never be done, and which might not support the AEC's predetermined solutions if it ever was done.

Fields clarified the AEC's attitude toward uncertainty in his discussion of the difference between the "viewpoint" of the AEC and that of the ACRS. Some members of the ACRS with a *"conservative viewpoint"* were worried that the "*gaps* which show up only when the reactor is put in operation may involve *unpleasant surprises* which would affect the safety of the plant."[11] These members wanted more data from experimental fast reactors. Such fears did not trouble the AEC staff. According to Fields, the staff "accepted the opinion that the *gaps in knowledge* which are discovered in the course of design can be corrected by design changes, and that those gaps in knowledge which are encountered only when the reactor is put into operation have *negligible probability* of making the reactor unsafe with containment."[12]

The AEC's assurances did not prevent three international labor unions with members living in the Detroit area from intervening in the Fermi licensing proceedings, making the Fermi plant the AEC's first contested reactor.[13] The AEC fought the unions all the way to the Supreme Court, which ruled in 1961 in favor of the AEC's right to issue construction permits for reactors with *unresolved safety problems*. The PRDC, which had continued construction while the case went through the courts, applied for an operating permit in 1961;[14] the plant began generating electricity in August, 1966. In October, 1966, the Fermi reactor suffered a *partial core meltdown* when a piece of zirconium metal, hastily added at the end of construction and not shown on the "*as built*" *drawings*, tore loose and blocked the coolant flow to part of the core.[15] The plant was briefly restored to operation in 1970, running mostly at low power levels as a research and development facility. In late 1972, it was announced that the Fermi plant would be dismantled,[16] never having come close to fulfilling the promise which the AEC envisioned in 1956.

The AEC was unable to keep the public from learning about the safety problems of the Fermi fast breeder reactor. But in 1958, while the Fermi case was still being litigated, the U.S. government participated in a coverup of the worst nuclear accident in history. This accident took place in the winter of 1957–58 in the southern Urals of the Soviet Union at a military plutonium production facility. An explosion of unknown origin heavily contaminated a large area with nuclear wastes. An unknown number of people were killed and injured, and at least thirty towns were subsequently erased from the map.

The cause of this accident and its full dimensions are still hidden by official secrecy. Both the Soviet and the U.S. governments covered up the accident, even though the Central Intelligence Agency had reports on file of the accident by the spring of 1958. The CIA was forced to release some of its documents in 1977, but is still hiding others in the name of national security.

News of the accident would have been embarrassing to nuclear developers in both nations. There was a vigorous worldwide movement against both countries' testing of nuclear weapons, and the report of a major weapons-connected accident would only have added to the strength of this movement. In the United States, knowledge of the disaster would have undercut the claims of nuclear developers that they could guarantee the safety of the entire nuclear fuel cycle. Even after the Urals accident became known in the United States in 1976, some nuclear developers tried to deny that any accident had occurred.

Zhores Medvedev, a Russian emigré and an internationally respected biologist, brought the Urals accident to public attention in an article on Russian dissident scientists that appeared in the November 4, 1976 issue of the British journal *New Scientist*. Medvedev, who did not know that the accident was unknown to his readers, wrote that a "tragic catastrophe" had occurred in the Urals when an explosion at a nuclear waste storage site spread radioactive dust over hundreds of square miles. Tens of thousands of people suffered from radiation sickness, and hundreds had died. The area was still closed to the public, and biologists were studying the effects on plants and animals.[17]

Neither the Soviet nor the U.S. government offered any response to Medvedev's claim. But the chairman of the British Atomic Energy Authority, Sir John Hill, immediately denounced Medvedev's account as *"rubbish," "pure science fiction,"* and *"a figment of the imagination."*[18] Having assumed the story was common knowledge, Medvedev was not prepared to defend himself against such strong attack. But another Russian emigré scientist, Professor Leo Tumerman, soon added some corroborating details. Tumerman said that during a 1960 car trip through the Urals, he passed through a large uninhabited area where highway signs told motorists not to stop for the next 20 miles and to drive at maximum speed because of radiation.[19]

In November of 1977, a Freedom of Information Act request forced the CIA to reveal for the first time that the U.S. government had known of the accident since the spring of 1958. The CIA documents reported that the occurrence of an accident in the Urals was widely known inside the USSR;[20] milk, meat, and other foodstuffs had been contaminated;[21] "several score" of people had died; and some villages "had been contaminated and burned down, and the inhabitants moved into new ones built by the government. They were allowed to take with them only the clothes in which they were dressed."[22] These documents were *sanitized*, some with entire pages blanked out. The CIA said that there "were also a number of documents which could not be released, even with deletions."[23]

Medvedev further bolstered his case by conducting a detailed review of the Soviet open literature on ecological studies which used radioactive materials to map out the complex relationships between plants and animals in a given area. He found numerous radioecological studies at unspecified sites where much higher levels of radioactive materials had been used than would normally be employed in field studies. The descriptions of the plants, animals, and soil types in these studies placed their location in the southern Urals.[24]

Simplified Diagram of a Pressurized-Water Reactor

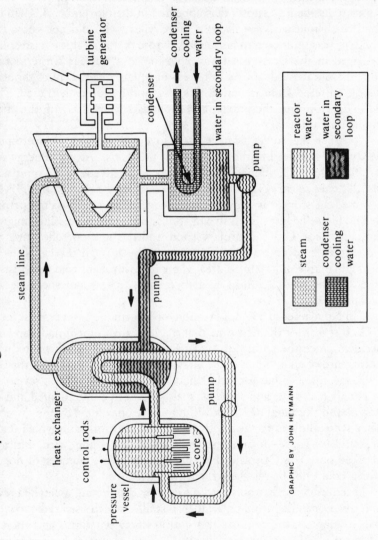

GRAPHIC BY JOHN HEYMANN

FIGURE 9 The nuclear reactor in a commercial nuclear plant heats water. The steam thus produced turns a turbine which runs a dynamo, generating electricity in the same manner as in a fossil-fuel-fired power plant. The commercial nuclear plants in the United States are of a type known as light-water reactors because they use water for cooling and to moderate—or slow down—the neutrons produced by the fissioning uranium atoms.

At the heart of the reactor is the reactor core, which holds the fuel rods. Uranium oxide is molded into cylindrical pellets about an inch tall and less than a half-inch in diameter. The pellets are stacked end-on-end in the fuel rods, thin tubes about 12 feet tall made of Zircalloy, an alloy of zirconium. This alloy shell is called the cladding. At a typical large nuclear plant, the reactor core consists of about 36,000 fuel rods, containing about 100 tons of uranium. The rods are arranged vertically in the core with space between them for cooling water to flow and for control rods, which are used to control the rate of fissioning, or to shut down the chain reaction entirely.

Control rods contain such materials as boron or cadmium, which absorb neutrons, *poisoning*—or inhibiting—the fission reaction. When the control rods are all inserted into the reactor core, they absorb so many neutrons that a chain reaction is impossible. To start a chain reaction, the control rods are slowly withdrawn. Plant operators control the amount of power the reactor produces by varying the distance to which the control rods are withdrawn. In an emergency, the control rods can be dropped rapidly into the core to halt the chain reaction, a process called a *scram*.

There are two types of light-water reactors: pressurized-water reactors (PWRs) and boiling-water reactors (BWRs). In a pressurized-water reactor, as shown at left, water flows through a closed system of pipes called the primary loop, passing over the reactor core. Water in the primary loop is kept under high pressure to keep it liquid, and is forced through the loop by large pumps. As it flows around the fuel rods, the water picks up heat. Then it travels through giant pipes, about 3 feet in diameter, to the heat exchanger.

In the heat exchanger, which functions like a giant radiator, heat is transferred from the primary loop to the secondary loop. Water in the secondary loop flashes into steam and is carried to the steam turbine, which runs the generator. The system is designed so that any radioactivity picked up in the water of the primary loop is kept isolated from the water in the secondary loop.

The water from the primary loop, which has now passed some of its heat to the water in the secondary loop, is pumped back into the reactor to cool the fuel rods again, to pick up heat, and to repeat its cycle.

The water from the secondary loop, which was turned to steam to drive the turbine, passes through devices called condensers, where it is condensed back into liquid and forced back to the heat exchanger again. The condensers are cooled by a third loop. In some reactors, water to cool the condensers is pumped from a body of water such as an ocean, river, or lake, forced over the condensers, and returned to the body of water. Other reactors use cooling towers, where the condenser water cascades along a series of steps inside a large cooling tower, releasing its heat into the atmosphere before being pumped back to the condenser.

In a boiling-water reactor, the water that flows over the core is allowed to boil in the reactor vessel, and goes on to turn the turbine directly, without the need for a secondary loop. Condensers cool the steam, and the liquid water is pumped back to the reactor core for another cycle.

While a nuclear explosion cannot occur in a light-water reactor, LWRs contain a tremendous amount of radioactive material, and great care must be taken to prevent a catastrophic release of these materials. Reactor designers use a philosophy

Medvedev's new evidence and the CIA documents strengthened his argument. Some nuclear developers were worried by the public-relations impact of Medvedev's disclosure. When Medvedev debated Edward Teller at Los Alamos in 1978, a scientist in the audience warned Medvedev, who is personally in favor of nuclear power, that his work would be used by people who opposed nuclear power: "There is a group of people in this nation determined to end the use of nuclear power and using the waste issue, they would use your words as ammunition."[25]

In October of 1979, three Los Alamos National Laboratory staff scientists and Harold Agnew, the president of General Atomic Corporation and a former director of Los Alamos, published an attack on Medvedev's hypothesis in an article in *Science* entitled "Are Portions of the Urals Really Contaminated?" This article, which cited only ten references (and six of those were Medvedev's own work and the CIA documents), made no mention of any of the Soviet radioecology literature which Medvedev had reviewed.[26]

Based on this cursory review, the Los Alamos group concluded that fallout from a Soviet nuclear weapons test was a *"more plausible"* explanation for any radioactive contamination that might have occurred than Med-

---

called *defense-in-depth* to try to guarantee safety. Defense-in-depth relies on *redundant safety systems*—or backup systems—to provide multiple levels of protection.

Three basic physical barriers have been designed to prevent the escape of radioactive materials from reactors. The first is the fuel rods themselves, which trap radioactive materials produced within their cladding. The second barrier is the reactor vessel, typically a 40-foot-high steel tank with walls 8½ inches thick. This tank is surrounded by two separate concrete and steel shields. The third barrier is the containment building that houses the reactor vessel; it is a reinforced-concrete structure with walls about 4 feet thick.

Even with these precautions, the risk of accidents that could release large amounts of radioactivity remains. To prevent the core from overheating, water must be constantly forced through the reactor vessel. Even after the reactor has been scrammed, it must continue to be cooled, because the core is so radioactive that the decay heat produced by the fission products is sufficient to cause the core to overheat. If cooling is interrupted—by a pipe break or a valve failure, for example—a *loss-of-coolant accident (LOCA)* would occur. Overheating could cause the fuel-rod claddings to melt. Further overheating could cause a *core meltdown,* an accident in which the fuel rods melt. If a sizable portion of the fuel should melt, the molten mass could then melt through the bottom of the reactor vessel, releasing large amounts of radioactivity into the containment building. The mass might even melt through the bottom of the containment building, producing what is called the *China syndrome,* potentially releasing large amounts of radioactive materials into the environment.

One important backup system is the *emergency core cooling system (ECCS),* which is designed to guard against core meltdowns. In the event of a loss-of-coolant accident, the ECCS is supposed to automatically flood the core with water, preventing the fuel rods from overheating.

vedev's suggestion that an accident in the handling of military waste might be responsible.[27]

Two months later, a team of scientists based at the Oak Ridge National Laboratory published a more detailed review of the Urals accident which demolished the claims of the Los Alamos report. The Oak Ridge study had 147 references, including an extensive survey of the Soviet radioecological literature. The authors, while initially skeptical of Medvedev's arguments, concluded that there had been "a *major airborne release* involving moderate- to long-lived fission products . . . in the winter of 1957 to 1958."[28] They also found that "the most likely cause of the airborne contamination was the chemical explosion of high-level radioactive wastes associated with a Soviet military plutonium production site."[29]

The Oak Ridge group rejected the Los Alamos group's suggestion that fallout from the explosion of a nuclear weapon might be responsible, because the ratio between the levels of strontium-90 and cesium-137 reported in the ecology studies could not have been produced by detonation of a nuclear weapon.[30] If the Los Alamos group had looked at these data, they could not have chosen fallout as their favored explanation.

The Oak Ridge team found further support for Medvedev's hypothesis by comparing maps of the region made before and after 1957. The names of more than thirty communities had been deleted from the post-1957 maps in the area Medvedev had identified, while populations were increasing in the rest of the region. In no other area of the region had so many communities disappeared from the maps.[31]

The Oak Ridge study ended with a plea to Soviet scientists to "share all pertinent information with others concerned with achieving the safe development of nuclear energy. Soviet experience gained during the application of remedial measures on an unparalleled scale following this accident is clearly unique and would be invaluable to the world nuclear community."[32]

By 1965, the public was growing more concerned about the hazards of nuclear accidents. Since the Fermi case and the publication of the Brookhaven Report (WASH-740), the industry had encountered several successful challenges to proposed reactors. One early fight took place in New York City after Consolidated Edison proposed building a 1,000-megawatt reactor in Ravenswood, Queens, just across the river from Manhattan.[33] One of the most well-known opponents of this reactor was David Lilienthal, the first chairman of the AEC, who left the commission in 1950. Lilienthal spoke against the Con Ed proposal at a 1963 meeting of the American Nuclear Society in New York. He later wrote that after his speech, "an AEC commissioner called a rump press conference in the lobby, and AEC employees, referring to themselves as a *'truth squad,'* set out to rebut my speech to the members of the press who were present."[34] Public opposition and lack of support from the AEC finally forced Con Ed to abandon the Ravenswood plant. In California, stiff citizen opposition forced Pacific Gas and Electric to cancel its plan for a reactor at Bodega Head, 50 miles north of San Francisco and close to the San Andreas fault.[35]

The last thing nuclear developers wanted to see was any new information that would make siting nuclear plants more difficult. AEC internal documents show that in 1965, the AEC suppressed a study of the consequences of reactor accidents for fear of its public-relations impact. In the summer of 1964, the AEC began an update of WASH-740. The 1957 report by scientists at the Brookhaven National Laboratory, which predicted that an accident at a small nuclear plant could cause 3,400 deaths, 43,000 injuries, and $7 billion in property damages, had been a thorn in the side of utility public relations ever since.[36] The AEC expected that the update, based on more operating experience than was available when WASH-740 was written, would show that the much larger reactors planned during the 1960s were safer than the plants examined in WASH-740. (The plant Consolidated Edison wanted to build in New York was five to ten times as large as the plant in WASH-740.) But when the update indicated that larger plants might produce more damage than that predicted in WASH-740, the AEC suppressed it—until a Freedom of Information Act request forced its disclosure in 1973.

Documents released in 1973 show that AEC officials hoped that the update would lower the WASH-740 estimates: "A major reason for reconsidering WASH-740 was that many people feel that new estimates would be lower."[37] Some people hoped that "new information" would reduce "some of the *pessimism* reflected in the WASH-740 hypothetical models. . . ."[38] But one memorandum warned:

> Great care should be exercised in any revision to avoid establishing and/or reinforcing the current popular notion that reactors are unsafe. Though this is a public information or *promotional problem* the AEC now faces with less than desirable success, I feel that by calculating the consequences of hypothetical accidents, the AEC should not place itself in the position of making the location of reactors near urban areas nearly indefensible.[39]

The AEC Steering Committee conducting the review soon learned that the update might cause public-relations problems. The Brookhaven scientists performing the study told the Steering Committee that the consequences of a core meltdown were "frightening," and that such an accident might affect an area "*equal to that of the state of Pennsylvania.*"[40]

Memoranda show that the Steering Committee was worried that "these results would strengthen opposition to further nuclear power."[41] Another memo cautioned that the impact of publishing the revised WASH-740 report on the reactor industry should be "weighed before publication."[42]

The commercial nuclear industry was also very concerned about the possible impact of the WASH-740 update. This concern was detailed in a speech called "Public Acceptance of Nuclear Power" by Pacific Gas and Electric public-relations officer Hal Stroube at a nuclear industry conference in 1965. Stroube had been fighting a losing battle with opponents of siting a PG&E plant at Bodega Head, California, and he wanted the AEC to "make it easier for us to get on with our chore of *public understanding*."[43] According to Stroube, the AEC could do the nuclear industry

a *great* [all emphases original] big favor by *cancelling* the now-in-progress updating of the Brookhaven Report. I've eaten a *steady diet* of WASH-740 in the past three years as it became the *bible of the anti-Bodega crowd*, and no amount of pointing to the disclaimers in the three page covering letter in front of the report could overcome those chilling words "3400 people killed, 43,000 injuries, $7 billion in property damage" which rolled so glibly off their tongues from the depths of this theoretical study. . . .[44]

Stroube's wish was granted. The AEC did not make any of the documents involved in the WASH-740 update public, and no technical report was published. As AEC Chairman Glenn Seaborg later said, "We didn't want to publish it because we thought it would be *misunderstood* by the public."[45]

The suppression of the WASH-740 update was only one of many attempts to hide the hazards of nuclear reactors by using information-management techniques. (In the same speech in which he proposed the cancellation of the update, Hal Stroube also suggested that the AEC eliminate the word *hazards* from the title of one of its reports.)[46] By the end of 1967, the AEC appeared to be well on its way to realizing its 1962 forecast of 1,000 nuclear plants in operation in the United States by the end of the century. Utilities had ordered seventy-five nuclear power plants,[47] and almost no one doubted that hundreds more would follow.

But as more people were confronted with nuclear plants in their area, doubts about safety began to grow, particularly about the effectiveness of the *emergency core cooling systems (ECCS)* which the AEC had begun requiring on all reactors in 1966. Despite the importance of the ECCS in preventing a core meltdown in the event of a *loss-of-coolant accident (LOCA)*, the AEC did not do even a small-scale test of an ECCS during the 1960s. The AEC had started building an experimental test reactor called *LOFT*, for *loss-of-fluid test*, in 1963, but this reactor was still not complete. In the absence of experimental data, the AEC claimed that the reliability of the ECCS was backed up by computer programs, or codes, whose *conservative* assumptions supposedly predicted the behavior of an ECCS during a loss-of-coolant accident.

By 1970, there were grave doubts within the AEC as to whether these computer codes adequately described what would happen during a loss-of-coolant accident. These doubts were greatly strengthened in late 1970 when a test of an ECCS, using a tiny, 9-inch-diameter simulated reactor core, resulted in the complete failure of the ECCS to deliver cooling water, a failure that was not predicted by the computer codes.[48]

Under fire from critics outside the agency who had learned of the test failure, the AEC backpedaled frantically. A special task force headed by Stephen Hanauer was convened, and on June 29, 1971, the AEC issued *Interim Acceptance Criteria* for ECCSs that the commission claimed provided *"reasonable assurance* that such systems will be effective in the *unlikely event* of a loss-of-coolant accident."[49] These criteria imposed little or no hardship on owners of reactors already in operation or under construction.[50]

The Union of Concerned Scientists, a nonprofit organization of scientists, engineers, and other professionals who have often taken issue with positions of the government on nuclear development, soon issued a strong attack on the adequacy of the criteria.[51] Intervenors against nuclear plants around the country began raising the issue in individual plant hearings. These events forced the AEC to hold a special hearing on ECCSs, which began in January, 1972. This hearing, which eventually produced 22,000 pages of testimony, exposed a seething mass of dissent within the AEC, shattering the AEC's calm public facade. The hearing demonstrated that the AEC had used a wide variety of information-management techniques to prevent the public from understanding how much uncertainty lay beneath its *"reasonable assurance"* of safety. During the course of the hearing, the public learned that:

☐ The AEC tried to intimidate members of its staff who were testifying at the hearing, passing out a list of *"Hints to AEC Witnesses."* One hint instructed commission employees: "Never disagree with established policy."[52]

☐ The AEC and its contractors punitively reassigned employees who publicly disagreed with established policy.

☐ The AEC censored public reports from its safety research laboratories.

☐ The AEC tried to prevent documents that contradicted its policy from being introduced into the hearings. Invoking *executive privilege*, the AEC denied intervenors the power of subpoena and the power of discovery, making it much more difficult for them to learn about information which the commission wanted kept secret.

☐ The AEC applied the principle of compartmentalization to keep information about safety research from some of its own staff. Documents from the safety labs were withheld from the regulatory staff. Safety researchers were instructed not to talk to the regulatory staff except under carefully monitored conditions.

The AEC's intimidation of its employees was documented by Robert Gillette of *Science*, who investigated the state of the AEC's safety research program while the ECCS hearings were going on in 1972. Gillette found that almost no one would talk to him on the record for fear of losing his job. At the AEC's research facilities in Idaho Falls, Idaho, Charles Leeper, president of Aerojet Nuclear Corporation (the contractor operating the site), was completely open about how Aerojet viewed any expression of dissent from the AEC's official line. Leeper said he had told the staff that they were free to tell Gillette their personal or professional opinions, but that if any employee's comments "sour his relationship with the customer [AEC], we cannot guarantee that after some time has elapsed that he will still be in his same position. We would, however, make every effort to find him a *suitable opening* in this organization, or elsewhere in Aerojet, or *allow* him to look beyond the company."[53] Aerojet employees were understandably unwilling to meet with Gillette in the company's offices, but several arranged backstreet meetings at night followed by drives to homes outside the town.[54]

Even criticism within the bureaucratic channels of the AEC could lead to action against an employee. One of the most important critics of the ECCS criteria was Morris Rosen, a technical advisor to the Director of Reactor Licensing. Rosen headed the Regulatory Staff branch of the AEC which handled ECCS performance evaluations from 1967 to January, 1972. In June, 1971, Rosen and an engineer who worked for him filed a memorandum that was highly critical of the Interim Criteria for ECCSs which the AEC was to make public later that month. They pointed out that the computer codes were still evolving, and that "it is not clear that the systems and the analytical tools are or will be *adequately understood* in the near future."[55] The memo concluded that "the consummate message in the accumulated code outputs is that the system performance cannot be defined with *sufficient assurance* to provide a clear basis for licensing."[56]

The AEC completely ignored this criticism. For his trouble, Rosen found himself reassigned to another position in January, 1972. The AEC said that the transfer was a promotion to a higher position with more responsibility. But Rosen was also moved away from his involvement with emergency core cooling. He told a reporter, "It's the sort of thing that, if it happened very often in an organization, you'd have to wonder."[57]

Despite his job change, Rosen testified in April, 1972 at the ECCS hearings. He said he was disturbed and discouraged "to continue to see the advice of what I believe can be considered a significant portion of, more likely, a majority of the knowledgeable people available to the Regulatory staff, still being basically disregarded." Rosen said that the regulatory staff did not have "knowledge sufficiently adequate to make licensing decisions for the . . . reactors operating or under construction." As to the AEC's analyses of the ECCS, Rosen said that they were "based on numerous unrealistic models, suffer from a number of restrictive assumptions and lack applicable experimental verification. Under these conditions . . . the effectiveness of emergency core cooling systems cannot be established."[58]

The special board conducting the ECCS hearings tried to keep the AEC's internal memoranda out of the hearing record by invoking *executive privilege*.[59] The AEC's staff counsel at the hearing claimed that the documents in question were exempt under the Freedom of Information Act, and the board agreed. The board said that withholding these documents "in no way prejudices the ability of participants in this proceeding to question the staff on the basis of its *technical conclusions*."[60] The board argued that revealing the documents would interfere with "the *open and frank exchange* of views and opinions among Government personnel who are responsible for the formulation of conclusions, frequently in areas where there may be differences of opinion."[61] After the threat of a lawsuit to be brought under the Freedom of Information Act by the intervenors challenging the Interim Criteria, the AEC overruled the hearing board and grudgingly ordered some of the documents released. The AEC said that "the documents contain, in the main, a mixture of data and opinion—in many instances, inextricably interwoven; and the bulk of their contents are, as the staff contends and the Board held, properly exempt from automatic production. . . . At the same time, they may provide a basis for fuller exposition of the staff testimony in the instant proceedings."[62] These documents proved to be invaluable to the intervenors in fully buttressing their case that there was extensive disagreement within the AEC about the ECCS criteria.

The hearings showed that the AEC's Division of Reactor Development and Technology (RDT) had been censoring public reports on safety research from Aerojet. On the witness stand, Milton Shaw, the director of RDT, admitted that the Aerojet reports were being censored: "*Censoring?* If you want to use that terminology in the sense I think you are using it, yes. . . . I think it is a basic requirement that reports that are issued by people who are working for us have in them *factual information.* . . ."[63]

The following day, J. Curtis Haire, the manager of Aerojet's nuclear safety program, was asked whether Shaw's censoring was "not a disagreement with . . . technical judgement, but, rather . . . an inhibition of a free and open discussion of [Aerojet's] views on safety?" Haire replied, "Yes, it is rather an inhibition of free and open discussion rather than a matter of taking issue with *technical matters.* . . . I believe that RDT is trying to avoid the problem or burden, if you will, of having to spend a lot of time answering public inquiries that are addressed to them."[64] Haire said that his belief was based on a conversation with Andrew Pressesky, Shaw's deputy for reactor safety. In keeping with the warning to his employees that Aerojet president Charles Leeper had given to *Science*, Haire was later removed from his job and given a position in charge of "program development."[65]

Reactor Development and Technology also kept the AEC's own regulatory staff in the dark about safety research that raised questions about the AEC's safety policies. For example, RDT did not give the regulatory staff a detailed Aerojet report on the areas where more research on ECCSs was needed. The report contained a table listing twenty-eight problem areas. In seven areas, techniques for analyzing the problem were *"missing."* The remaining techniques were placed in one of five categories: *"Preliminary," "Unverified," "Inadequate," "Imprecise,"* and *"Uncertain."* In terms of research priorities, eight areas were classified as being *"Very High,"* and fifteen more as being *"High."*[66]

Reactor Development and Technology also told researchers at Oak Ridge and Idaho not to speak to members of the regulatory staff about various controversial topics in reactor safety. The only officially tolerated contacts were prearranged meetings in Washington which were chaperoned by officials from RDT. During these meetings, safety researchers were only allowed to answer specific questions from the regulatory staff and were not supposed to raise their own worries about the safety of specific nuclear plants.[67]

Long before the ECCS hearings, people had been calling for the AEC to do more safety research. Many reports from Oak Ridge and Idaho complained about the lack of experimental data to support the computer codes. One Aerojet report, for example, found that "reliable, applicable experimental data are desperately needed to evaluate and demonstrate analytical capabilities. . . ."[68] Almost every year the ACRS sent the AEC a letter asking for more safety research. In a February 1972 letter to AEC Chairman James Schlesinger, ACRS Chairman C. P. Siess complained that "there has not yet been formulated a sufficiently specific definition of the national safety research needs for water reactors, including the means and schedules to be used in resolving problems."[69] While AEC researchers were growing more concerned about the safety of the ECCS, the AEC was cutting the safety research budget.[70] In 1971, the AEC terminated a number of research projects that appeared to be on the verge of producing experimental results that might undermine the AEC's public position on safety.[71]

The AEC's main defense of its modest safety research program was its well-publicized philosophy of *conservatism:* whatever gaps there might be in its understanding of reactor safety could be covered by the appropriate *conservatisms* in engineering design and computer codes. In its written testimony presented at the beginning of the ECCS hearings, the AEC regulatory staff concluded that "the conservatism in today's criteria and today's evaluation models suffice to provide *reasonable assurance* of the safety of today's reactors."[72]

Here again, the hearing revealed that there were disagreements within the AEC about what the word *conservative* meant. George Lawson, a heat transfer specialist at Oak Ridge, described the difficulties he had in assessing how conservative the AEC's criteria were:

The assertion is that conservative assumptions are made where possible, and this is true. But there are some areas where, in my opinion, we don't know whether the assumption we are making is *conservative* or not because we don't know what is occurring physically.[73]

Lawson said that he was not sure

whether the conservatisms existing in other places in the calculations are sufficient to overcome the possible *unconservatisms* in certain assumptions.

Now in order to answer that question fully, one must have a cataloging of the conservatisms and their possible effect on the outcome of the accident. That *cataloging of conservatisms* does not exist.[74]

CHAPTER TEN

The Too Favorable Lesson

Nuclear developers have found that reassuring numbers—regardless of their accuracy—can be very useful in lending authority to soothing public-relations statements. Fuzzy numbers can be translated into crisp public-relations material and used to shape opinion and influence legislation. The case of the *Reactor Safety Study (RSS)*, the most famous study of the risks of nuclear reactors, provides a good example of the way in which questionable results can be converted into political propaganda.

The RSS is also known as *WASH-1400*, or the *Rasmussen Report*, after its director, Norman C. Rasmussen of the Massachusetts Institute of Technology's Nuclear Engineering Department. When work on the RSS began in 1972, public support for nuclear power was slipping, and the AEC (and later the NRC) and industry officials hoped the report would halt the decline. The goal of the RSS was to produce numerical estimates of the likelihood of reactor accidents. Nuclear developers expected that these estimates would put to rest the growing public concern about reactor safety by placing the risks and benefits of nuclear power in *perspective* with other societal risks.

A draft of the Rasmussen Report was released in 1974, and the study was completed in 1975. Its numerical estimates were very reassuring: an individual's chances of dying in a reactor accident were said to be 1 in 5,000,000,000.[1] The study's *Executive Summary* proclaimed that "the likelihood of reactor accidents is much smaller than that of non-nuclear accidents having similar consequences. All non-nuclear accidents examined in this study, including fires, explosions, toxic chemical releases, dam failures, airplane crashes, earthquakes, hurricanes and tornados, are much more likely to occur and can have similar consequences to, or larger than, those of nuclear accidents."[2] On its second page the summary displayed a graph which compared the risk of nuclear accidents with the risk from falling meteors. The graph showed these risks to be approximately equal.[3] In the press and in industry PR material, the conclusion of the fourteen-volume report was summed up in a single phrase: "the chance of *mass destruction* from an atomic reactor accident was as unlikely as that of a *meteor* striking an urban area."[4]

The Rasmussen numbers became the backbone of the nuclear developers' drive to build a consensus that nuclear power was safe. The results were used in ads, brochures, speeches, television appearances, and press releases. Its Executive Summary was widely distributed to Washington decision makers, the media, and other opinion makers. The report was used to lobby Congress. And it played a role in the defeat of antinuclear referenda in six states in 1976.

Throughout this PR blitz, little mention was made of the limitations of the report, even though flaws in the study were immediately apparent. Since then, the methods and data the RSS team used have been sharply criticized, and serious questions have been raised about the validity of the results. Most suspect were the numerical risk estimates—the most widely publicized conclusions of the study. Some critics of the Rasmussen Report maintain that it is in principle impossible to support numerical estimates of the risk of nuclear accidents with convincing theoretical arguments, saying that the likelihood of error in the safety analysis makes the results drawn from it meaningless.[5]

Even within the AEC, there were questions about the feasibility of developing numerical estimates of reactor risk. In 1973, while work on the RSS was in progress, the AEC completed a report that cast doubt on the possibility of accurately quantifying reactor risk. The report, known as the *Ernst Report*, after task force director Malcolm I. Ernst of the AEC's staff, concluded that it would be hard to estimate the risk of nuclear accidents "since identification of all possible accident combinations has not been accomplished." The report also pointed out that experience with operating reactors raised further questions about optimistic accident probability estimates.[6]

Because the Ernst Report might have been politically embarrassing, the AEC suppressed it—until a copy was leaked to the Union of Concerned Scientists in 1974. The AEC then released a censored version, deleting those conclusions which might have undermined the credibility of the Rasmussen Report.[7]

The risk estimates in the RSS were developed using a technique called *fault-tree analysis*. A fault tree begins with a failure and attempts to trace it back to its causes. To estimate the probability that a car will fail to start (see figure 10), for example, one begins by listing all of the events that could cause the car to fail, such as an insufficient battery charge, a defective ignition system, and so on. Then one lists all the events that could cause each of those events, and so on. When the fault tree has been mapped out in sufficient detail, specialists assign probabilities to each occurrence on each pathway. These probabilities are then combined to give an overall failure probability, which is a numerical estimate of the likelihood that the car will fail to start.

Starting with data for the failure rates of system components, such as valves, pipes, and reactor operators, the RSS team used fault trees to estimate the probability of various kinds of nuclear accidents. It then developed estimates of the probability of releases of different amounts of radioactive material. Combining these estimates with data on weather patterns and population density, it developed numerical estimates of the *risk to the public*.

The use of fault-tree analysis requires detailed knowledge about the equipment and systems being analyzed. In a nuclear reactor, the number of possible sequences of events which could lead to accidents is extremely large, perhaps on the order of billions.[8] While fault-tree analysis is theoretically workable, the problems one must face when attempting to use it for complex systems are staggering.

There has been tremendous controversy about whether fault-tree analysis is useful for estimating the risks posed by machines as complicated as nuclear reactors. Central to this debate is the issue of *completeness*. All fault trees, like the one for getting your car started, contain a category for *All Other Problems*. This category must be used because it is impossible to demonstrate that a fault tree is complete. The fault-tree analysts try to ensure that the All Other Problems category contains no *significant accident pathways*. But despite their best efforts, the question remains: what might have been left out?

Psychological data reported by Decision Research—a group of psychologists who study human decision making—suggest that people are not very good at recognizing errors in fault trees:

> People are quite insensitive to how much has been left out of a fault tree. Deleting branches responsible for about half of all automobile starting failures only produced a 7% increase in people's estimates of what was missing. Professional automobile mechanics were about as insensitive as nonexperts. Apparently, what was out of sight was also out of mind.[9]

The specialists at Decision Research argued that "despite an appearance of objectivity . . . ," fault-tree analysis was "inherently subjective."[10] Reliability analyst William Bryan agreed, pointing out that "everybody will draw a different fault tree. No two people will go through the same mechanism because you're making judgments at every branch. . . ."[11]

Among the judgments to which Bryan referred are decisions about which of the billions of possible accident pathways are sufficiently *credible* to justify including them in the fault tree. Those which do not seem credible are ignored.[12]

Another set of judgments centers on the selection of failure probabilities. Engineers have enough data on many reactor components to assess their reliability quite accurately. It is more difficult to estimate the likelihood of human error or to assess the probability of failure in situations involving processes that are not fully understood. In these cases engineers

Fault Tree Showing Ways that Car Could Fail to Start

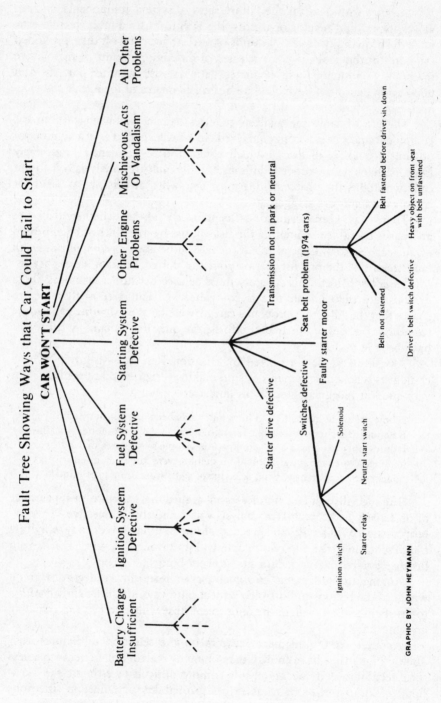

CAR WON'T START

Battery Charge Insufficient

Ignition System Defective

Fuel System Defective

Starting System Defective

Other Engine Problems

Mischievous Acts Or Vandalism

All Other Problems

Starter drive defective

Switches defective

Faulty starter motor

Seat belt problem (1974 cars)

Transmission not in park or neutral

Ignition switch

Starter relay

Solenoid

Neutral start switch

Belts not fastened

Driver's belt switch defective

Belt fastened before driver sits down

Heavy object on front seat with belt unfastened

GRAPHIC BY JOHN HEYMANN

must rely on *subjective probabilities* [13]—that is, educated guesses. The problem with such guesses is that the mindset of the assessor may bias the estimate in an optimistic or pessimistic manner. Psychological experiments have shown that people tend to underestimate risks when they believe that they have control over a situation. Most people, for example, think that they are among the safest drivers on the road, and that they are less likely to injure themselves with knives, power mowers, bicycles, and other hazardous products than the rest of society. [14]

In addition to the theoretical difficulties associated with fault-tree analysis, critics of WASH-1400 have pointed out many other flaws in the report:

☐ The data base used for the RSS was highly suspect and often inadequate. [15] Much of the data was supplied by the reactor manufacturers and was not independently checked. Computer models supplied by Westinghouse were not reviewed to determine their validity. [16]

☐ The RSS failed to include the possibility of sabotage in its analysis.

☐ The RSS did not take aging of reactor components into account. For this reason, the report admitted that its results should only be considered applicable to the five-year period immediately following completion of the report. Despite its obvious importance, this limitation was not highlighted in the report. [17]

☐ The treatment of *earthquake risk* was inadequate. Strong quakes, known in the industry as *seismic events*, could result in multiple or simultaneous ruptures in pipes or vessels, triggering serious accidents. [18]

☐ The Executive Summary and Main Report greatly underestimated the number of cancers, genetic effects, and other life-shortening effects that would result from a release of a large amount of radioactive material. [19]

☐ The study, a massive document 3,300 pages long in fourteen volumes, was poorly organized, making it difficult to follow closely the thread of extended technical arguments. The lack of an index compounded this problem.

Despite the uncertainty about the accuracy of the results of WASH-1400, the AEC and the nuclear industry almost never mentioned these limitations when communicating with the public. Instead, the press releases

FIGURE 10 A fault tree showing ways in which an automobile might fail to start. Deletion of branches responsible for about half of all failures produced only a 7% increase in people's estimates of what was missing. Professional auto mechanics were about as insensitive to changes as ordinary drivers. In contrast to an auto, which is a relatively simple system, a nuclear reactor has thousands and thousands of components, making fault-tree analysis far more difficult.

Source: Adapted from Baruch Fishhoff, Paul Slovic, and Sarah Lichtenstein, "Fault trees: sensitivity of estimated failure probabilities to problem representation," *Journal of Experimental Psychology: Human Perception and Performance*, 1978.

and public-relations material emphasized the conclusion that the risk from nuclear power reactors was negligible, without qualification.

When the draft of the study was completed in August, 1974, the AEC heralded its conclusions with a press release bearing the heading, "RISK TO PUBLIC FROM NUCLEAR POWER PLANTS VERY SMALL, STUDY CONCLUDES." The release described the RSS as a "two-year *independent* study of nuclear reactor safety—the most *definitive* ever undertaken." The credibility of the report was further boosted by a reference to its "technical staff of about 60 scientists and engineers, plus a large number of specialized consultants."[20] At no point in the news release did the AEC mention any of the limitations of the study, even though a number of problems had already been identified by a group of twelve AEC officials and outside consultants who had given the draft report a hasty, two-week review.[21]

Actually, many of the limitations and uncertainties of the report were acknowledged in its appendices, but these disclaimers did not make it into documents designed to be read by the general public. Commenting on this process of selective editing, the Union of Concerned Scientists noted that

> as one moves from the very technical material (mostly in the RSS appendices) to the assembly and explanation of the study's results (in the Main Report) to the publicly disseminated summary (the 12 page Executive Summary), to the nuclear industry's and the reactor owner's interpretation of the RSS into soothing advertisements for nuclear power (clean, safe, and cheap), the cautionary notes, the uncertainties, the sense that there are important limitations to the RSS results ... all these successively drop away.[22]

The UCS pointed out that "these vanishing cautions" were replaced by "increasingly forceful declarations that nuclear technology is benign and that this conclusion is, at last, overwhelmingly confirmed."[23]

Misleading editing is particularly apparent in the Executive Summary. For example, the study was based on analysis of only two existing plants, one pressurized-water reactor and one boiling-water reactor. The Main Report concedes that "the differences in design between various plants" raise questions about how applicable these results are to nuclear plants in general.[24] The Executive Summary, however, gives a very different picture of the scope and applicability of WASH-1400, not mentioning that only two reactors were studied. Instead it says that "the study considered large power reactors of the pressurized water and boiling water type being used in the U.S. today,"[25] thus implying that the results apply to all of the large reactors in the country.

Because the Executive Summary was much shorter than the Main Report or its appendices, it was by far the most widely read part of the document. In addition, the AEC distributed copies of the summary free of charge, while charging $174.50 for a paper copy of the complete report and its appendices.[26] Because few members of the public and few policy makers had the time or the desire to wade through the complete 3,300-page report, the result of selectively editing the summary was to mislead people about the contents of the report.

When WASH-1400 first came out, media coverage included little that would indicate that the RSS results were anything less than clearly proved scientific *fact*. Members of the media rarely have the time or patience to spend days or weeks poring over hard-to-follow, poorly organized, technical reports. They rely primarily on news releases and concise explanations like the Executive Summary. Media coverage of the RSS results read like an AEC news release; the limitations of the study were rarely mentioned.

When the draft report was released in August 1974, media coverage centered on the comparison made in the Executive Summary that a serious nuclear accident is about as unlikely as the destruction of a city by a meteor. The *New York Times* article on the draft report trumpeted forth the meteor message in its first paragraph: "The Atomic Energy Commission released today a study of nuclear power plant safety indicating that the chance of mass destruction from an atomic reactor accident was as unlikely as that of a *meteor* striking an urban area—one in a million years."[27] The *Times* article did not discuss the limitations of the study, nor did it mention its omission of the possibility of sabotage, which was briefly noted in some newspapers. Several days later, the *Times* announced that even in light of the assurances of safety the RSS had provided, the AEC had no plans to ease safety requirements at nuclear plants.[28]

The *Wall Street Journal*, in an editorial published the day after news of the RSS hit the press, applauded the study for putting "the crux of the nuclear argument into *realistic perspective*." The editorial went on to make the meteor comparison and, after mentioning the sabotage caveat, concluded that "the U.S. had *no choice* but to press forward with nuclear energy."[29]

Time magazine showed more political sophistication in its coverage than most other major media outlets, saying that "for the better part of two decades" the AEC had been "bedeviled by a problem of its own making"—WASH-740. "Last week," *Time* reported, "the AEC sought to put the old report to rest forever by issuing WASH-1400." After a quick rendition of the meteor comparison, the article pointed out that the report "does not even consider sabotage," and said that the UCS planned to scrutinize the report for "any oversimplification" or error which might call the results into question.[30]

When the final version of the RSS was completed in October, 1975, the newly formed NRC—in the tradition of the AEC—issued another reassuring news release. The release quoted NRC Chairman William A. Anders as saying that the RSS provided "an *objective* and *meaningful* estimate of the public risks" of present-day nuclear reactors. "The final report is a *soundly based* and *impressive* work," said Anders. "Its overall conclusion is that the risk attached to the operation of nuclear power plants is very low when compared with other natural and man-made risks." The news release made no mention of the limitations of the study.[31]

A number of careful critiques of WASH-1400 brought its limitations out into the open. In 1977, for example, the Union of Concerned Scientists published a 200-page book pointing out the weaknesses of the RSS.[32] Slowly, the Rasmussen Report began to be known as the *controversial* Rasmussen Report. Finally, those critics who charged that the methods used by the RSS team could not provide accurate numerical estimates of reactor risk were vindicated when, in 1979, the NRC withdrew its support for the risk estimates.

The decision by the NRC to stop using the risk estimates from WASH-1400 was prompted by the report of the Risk Assessment Review Group, chaired by Professor Harold Lewis of the University of California at Santa Barbara, and widely known as the *Lewis Report*. The Lewis Report was the final product of a year-long review of the RSS. Its effect on the credibility of the Rasmussen Report was devastating.

The Risk Assessment Review Group was established by the NRC in 1977 at the request of Representative Morris Udall. Over the next year, the review group, which included scientists and engineers from the University of California, Princeton University, the Environmental Protection Agency, the Brookhaven National Laboratory, the Electric Power Research Institute, and the California Institute of Technology, met twelve times. The group's report was completed in September, 1978, and was released promptly.[33]

The Lewis Report concluded that "WASH-1400 was a substantial advance over previous attempts to estimate the risks of the nuclear option."[34] But the review group also had some harsh words for the study:

WASH-1400 is defective in many important ways. Many of the calculations are deficient when subjected to careful and probing analysis, with the result that the accuracy of many of the absolute probabilities calculated therein is not as good as is claimed.[35]

The statistical analysis in WASH-1400 leaves much to be desired. It suffers from a spectrum of problems, ranging from the lack of data on which to base input distributions to the invention and use of wrong statistical methods. Even when the analysis is done correctly, it is often presented in so murky a way as to be very hard to decipher.[36]

We are unconvinced of the correctness of the WASH-1400 conclusion that they [such events as fires, earthquakes, and *"human accident initiation"* (i.e. sabotage)] contribute negligibly to the overall risk.[37]

One key deficiency is the use by the study team of some methodological and statistical assumptions that lack credibility. Therefore, the absolute values of the risks presented by the Report should not be used uncritically either in the regulatory process or for public policy purposes.[38]

In addition to these technical criticisms, the Lewis Report criticized the way the RSS was presented and used:

We find that the Executive Summary is a poor description of the contents of the report, should not be portrayed as such, and had lent itself to misuse in the discussion of reactor risks.[39]

There are instances in which WASH-1400 has been misused as a vehicle to judge the acceptability of reactor risks. In other cases it may have been used prematurely as an estimate of the absolute risk of reactor accidents without full realization of the wide band of uncertainties involved. Such use should be discouraged.[40]

The Lewis Report led the NRC to take seriously for the first time the issues raised by critics of the Rasmussen Report. In January, 1979, the NRC "reexamined its views" on the RSS, announcing that "the Commission does not regard as reliable the Reactor Safety Study's numerical estimate of the overall risk of reactor accident." In addition, the commission admitted that the peer review process used in connection with WASH-1400 was "inadequate" and withdrew "any explicit or implicit endorsement of the Executive Summary."[41]

Some nuclear developers, including Norman Rasmussen himself, charge that the NRC misunderstood the message of the Lewis Report and therefore set out to discredit the entire RSS.[42] NRC Commissioner Peter Bradford disagrees with this charge, calling such interpretations "revisionist versions of . . . history."[43] Bradford says that "WASH-1400 was substantially misused almost from the date of its completion by the NRC, and what resulted was a near classic pendulum swing from exuberance to disavowal."[44] As a result, Bradford maintains, the merits of the study became obscured in the controversies over the uses to which it was put.

Bradford thinks that the commission could have used the study to help identify weak spots in reactor design and operation that needed more regulatory attention. Instead, the NRC and the industry used the study "for propaganda purposes."[45]

"The fundamental point," according to Bradford, "is that confidence in risk assessment, like much else in nuclear regulation, was stated too positively too soon. As in other areas such as nonproliferation, waste management, and the harmful effect of low-level radiation, there was a political backlash to what had originally been a politically motivated optimism."[46]

Bradford believes that it is as important for the NRC and the nuclear industry to learn from the WASH-1400 experience as it is for them to learn from accidents like Brown's Ferry and Three Mile Island:

> WASH-1400 was not ... misapplied because the Commission did not understand its basic message or because the Commission is somehow undermanaged. WASH-1400 was misapplied because the Commission wanted to learn something other than what the data and the study were able to teach it.
>
> Here—as in past inattention to nonproliferation—as in the wildly optimistic AEC economic and demand forecasts of a decade ago—as at Lyons, Kansas ... as at St. George, Utah ... as in the failure to learn at least about emergency preparedness and communication with the control room in the wake of Brown's Ferry ... one sees again the desire among the regulators and in the industry to learn the wrong lesson, to broadcast a message more certain than a disciplined analysis of the data would support, to understate the uncertainties a great deal.[47]

Bradford says that politically motivated optimism has led to many of the nuclear industry's problems: "If one follows a fault tree from much of the friction and paralysis that today are said to affect the nuclear licensing process, one comes eventually for a root cause to the impulse to learn the *too favorable lesson* from any experience."[48]

CHAPTER ELEVEN

Business As Usual

The "root cause" of the accident at Three Mile Island—the most serious reactor accident in this nation's history—was the nuclear mindset. The President's Commission on the Accident at Three Mile Island (also known as the *Kemeny Commission* after its chairman John G. Kemeny) said that the word *mindset* occurred over and over again in testimony before the commission. The commission concluded that over the years, "the belief that nuclear power plants are *sufficiently safe* grew into a conviction. One must recognize this to understand why many key steps that could have prevented the accident at Three Mile Island were not taken. The Commission is convinced that this attitude must be changed to one that says nuclear power is by its very nature potentially dangerous, and, therefore, one must continually question whether the safeguards already in place are sufficient to prevent major accidents."[1]

The commission said that while popular discussions of nuclear plants tend to concentrate on questions of equipment safety, the accident at Three Mile Island showed that the "fundamental problems are *people-related problems*":[2] "our investigation has revealed problems with the 'system' that manufactures, operates, and regulates nuclear power plants. . . . Wherever we looked, we found problems with the human beings who operate the plant, with the management that runs the key organization, and with the agency that is charged with assuring the safety of nuclear power plants."[3]

The commission concluded that given the shortcomings of Metropolitan Edison (the utility that owns Three Mile Island), of the manufacturers of equipment, and of the Nuclear Regulatory Commission, "an accident like Three Mile Island was eventually inevitable."[4] And the Kemeny Commission warned that *"to prevent nuclear accidents as serious as Three Mile Island, fundamental changes will be necessary in the organization, procedures, and practices—and above all—in the attitudes of the Nuclear Regulatory Commission and, to the extent that the institutions we investigated are typical, of the nuclear industry* (original emphasis)."[5]

The Kemeny Commission's conclusions were buttressed by the other major study of the accident, which was sponsored by the NRC itself, and was performed by a Special Inquiry Group. It is known as the Rogovin Report, after the group's director Mitchell Rogovin. The Rogovin Report concluded that, before the accident,

> an *attitude of complacency* pervaded both the industry and the NRC, an attitude that the engineered design safeguards built into today's plants were more than adequate, that an accident like that at Three Mile Island would not occur—in the peculiar jargon of the industry, that such an accident was not a *"credible event."*[6]

The Rogovin Report also found that

> unless *fundamental* [original emphasis] changes . . . are made in the way commercial nuclear reactors are built, operated, and regulated in this country, similar accidents—perhaps with the potentially serious consequences to public health and safety that were only narrowly averted at Three Mile Island—are likely to recur.[7]

A long list of failures, most of them *people-related problems* that grew out of the nuclear mindset, contributed to the accident at Three Mile Island. These failures included operator error, poor operator training, poor reactor control-room design, failure of communication within and between the companies that built and managed the plant, failure of the regulatory process, and poor emergency planning.

During the initial stages of the accident at Three Mile Island, the reactor operators shut off the emergency core cooling system (ECCS), which had been automatically activated in response to the loss of coolant water caused by the failure of a valve to close. This turned out to be a serious error: if the operators had not shut off the ECCS, the accident would have remained minor. For this reason, many analyses of the accident at TMI have said that the cause of the accident was *operator error*.[8]

The Kemeny Commission objected to this conclusion, saying that while it was true that operator error contributed to the accident, there were a number of factors that led the operators to err. First, the commission said that the training of TMI operators was "greatly deficient."[9] The operators, who had been trained by the reactor manufacturer Babcock and Wilcox and licensed by the NRC, were "unprepared to deal with something as confusing as the circumstances in which they found themselves."[10] An important tool in the training of operators is the use of a *simulator*—a mock control-room console which is supposed to reproduce events that can happen in the reactor. But the simulator, designed by Babcock and Wilcox, was not programmed to reproduce conditions similar to those that occurred in the TMI accident, so the operators had no experience with those accident conditions.[11] The Kemeny Commission found that operator training was poor

PLATE E Novel modification to reactor control panel. Due to poor "human factors engineering," reactor operators are often faced with controls that have very different functions but look and feel identical. Workers at one plant replaced indistinguishable controls with those shown above.

Source: Electric Power Research Institute.

throughout the nuclear industry. The operators at Three Mile Island had scored higher than the national average on the NRC licensing examination.[12]

A second factor that led to operator error at Three Mile Island was the poor design of the control room (see plate E). The Kemeny Commission found that the "control room was not adequately designed with the management of an accident in mind."[13] In the early stages of the accident, more than 100 alarms went off in the control room, and the operators had no way of identifying the important ones and suppressing the unimportant ones. In the resulting confusion, it is hardly surprising that errors were made.

The poor design of the control room—or poor *human factors engineering*, as such design is known—provides a good example of failure of communication within the nuclear industry. The danger of having too many alarms was recognized during the design stage by Burns and Roe, the architect-engineering firm for the plant. Nevertheless, the control room was not redesigned.[14]

The failure to resolve the control-room design issue is not the only example of failures to solve known problems or heed warnings about dangers in plant procedures or design. A year and a half before the accident at Three Mile Island, the same kind of valve that failed to close at Three Mile Island stuck open at another Babcock and Wilcox reactor, the Davis-Besse plant. Operators interfered with the emergency core cooling system, just as they later did at TMI, but because the reactor was operating at only 9% power, a serious accident was averted.[15]

Both Babcock and Wilcox and the NRC investigators recognized the potential for a serious accident in the Davis-Besse *incident*. In a memo written more than a year before Three Mile Island, a Babcock and Wilcox engineer said that if the Davis-Besse failure had occurred in a reactor operating at full power, "it is quite possible, perhaps probable, that *core uncovery* and possible fuel damage would have occurred."[16] The warning "fell between the cracks."[17] B&W and the NCR failed to instruct utilities on the correct operator actions in case of a recurrence of the valve failure.[18] At TMI precisely what the B&W engineer had predicted, happened.

Pointing to examples like Davis-Besse, the Kemeny Commission concluded that in a number of important cases, Metropolitan Edison, Babcock and Wilcox, and the NRC "failed to acquire enough information about safety problems, failed to analyze adequately what information they did acquire, or failed to act on that information. Thus there was a serious lack of communication about several critical safety matters within and among" the parties involved in the building and operation of TMI.[19]

The Kemeny Commission also criticized the NRC's program of *Inspection and Enforcement (I&E)*, which is supposed to ensure that equipment and procedures in use at nuclear plants meet the agency's standards. The Kemeny Commission said there were "serious deficiencies" in the NRC's I&E program.[20] I&E inspectors did little independent checking of procedures and equipment, relying heavily on the utility's *self-evaluation*. An inspection at TMI conducted shortly before the accident consisted solely of examining company records and interviewing plant personnel.[21] The Kemeny Commission also pointed out that utilities and reactor manufacturers "often have a strong *financial disincentive* to evaluate and report safety problems that may result in more stringent regulations."[22]

Metropolitan Edison's *quality assurance plan* also came under attack from the Kemeny Commission, which noted that the utility did not employ enough inspectors to conduct the inspections called for under the plan.[23]

Furthermore, the Kemeny Commission criticized emergency planning at TMI, saying that it "had a low priority in the NRC and the AEC before it. There is evidence that the reasons for this included their confidence in designed reactor safeguards and their desire to avoid raising *public concern* about the safety of nuclear power."[24]

The problems and attitudes that led to Three Mile Island were not unique. On the contrary, long before the TMI accident there were indications that the institutions that build and run nuclear plants were not performing to specifications. The nuclear mindset, however, helped filter these warnings out, so they generally were not perceived, acknowledged, or acted upon by the people running the system.

One of these warnings was the *Nugget File*, a special internal file of problems at nuclear plants maintained by Stephen H. Hanauer, a senior safety official on the NRC staff. The Union of Concerned Scientists obtained the Nugget File under a Freedom of Information Act request. The file, a twelve-inch-thick stack of short reports, contained descriptions of some astonishing *occurrences*, some of them serious, some trivial, some ludicrous. Taken together, the reports in the Nugget File illustrate the complacent attitude of people in the nuclear power program. (In a marginal note on one of the reports, Hanauer wrote: "Some day we will all wake up.")

☐ At an unidentified reactor (the Nugget File did not give the name) in March 1968, workers used a basketball to plug a pipe. This makeshift seal was used when modifications were being made to the plant's spent fuel pool cooling system. Workers had to cut a pipe that was connected to the pool and relocate a pump. To prevent water from leaking out of the pool through the pipe, they wrapped a *"regulation basketball"* with two inches of rubber tape, inserted it into the pipe and inflated it. They then cut the pipe. Work was under way when the basketball was blasted through the pipe and out the open end. Fourteen thousand gallons of water followed closely behind, gushing into the basement. As a *"Corrective Action,"* a *"more conventional* seal . . . was substituted for the basketball." The AEC concluded that "where the risk of fuel melting and personnel safety are involved, consultation with knowledgeable people should be made prior to *questionable operation."*[25]

☐ At another unidentified reactor in April 1966, radioactive water was found in one of the plant drinking fountains. Investigators discovered a hose connected from a well-water tank to a 3,000-gallon radioactive waste tank. The AEC said that the "coupling of a contaminated system with a potable water system is considered *poor practice* in general. . . ."[26]

☐ After a relay failed in the safety circuits of the Prairie Island Unit 1 reactor in Red Wing, Minnesota in May 1975, investigation revealed that a coil of the wrong size had been installed. A mistake in the spare parts manual had led to the error. Eight other relays were found with improper coils.[27]

☐ At the Brown's Ferry Units 1 and 2 reactors in Decatur, Alabama in May 1976, part of the support systems that hold up the huge reactor pressure vessels were installed sideways. This was done in both reactors. As a "corrective action," new shims were installed, "thus *meeting the intent* of the original design." The *"event"* was attributed to *"personnel error"* during construction.[28]

In March 1975, the Brown's Ferry plants were the site of a fire that was one of the most serious reactor accidents in U. S. history. Although a catastrophe was averted, the fire ended up costing over $240 million for repairs and replacement power. Like the accident at Three Mile Island, the Brown's Ferry fire pointed to extensive problems in the management of nuclear power.

The fire began when a worker testing for air leaks with a candle accidentally ignited some flammable foam that was being used to seal the leaks. The fire spread quickly, burning the insulation on electrical cables that flowed out of the cable-spreading room under the reactor control room, which served both reactors. About 1,600 cables were damaged. After their insulation burned off, the cables short-circuited. The electrical controls for valves, pumps and blowers were wiped out, as well as instruments that fed information to the reactor operators in the control room. The emergency core cooling system for Unit 1 was rendered inoperable, as were portions of the Unit 2 ECCS.[29] Reactor operators struggled to manage the accident in a smoke-filled control room. They succeeded in shutting down both reactors and jury-rigging pumps to supply cooling water to the reactor cores. The fire was finally extinguished about seven hours after it began.

An NRC review group appointed to study the accident—which the review group generally called an *incident* or an *event*—wrote that the specific cause of the fire was an *"undesirable combination* of a highly combustible material . . . and an *unnecessary ignition source.* . . ."[30] The review group did reach some more illuminating conclusions about the cause of the accident, however, pointing to such factors as deficiencies in the plant's *quality assurance* program, poor firefighting and prevention capabilities, inadequate separation of cables, using flammable foam to seal air leaks, testing for leaks with a candle, and failure to pay attention to earlier candle-induced fires at the site.[31] In addition, the review group reported that there were *"weaknesses"* in emergency response plans in case of accident.[32]

Since the accident at Three Mile Island, there have been a number of warnings about *people-related problems* at nuclear plants. The *Boston Herald American* did an exposé in 1980 on drinking and drug abuse among maintenance workers at Boston Edison's Pilgrim plant. The *Herald* reported that workers injured on the job were intoxicated on alcohol and amphetamines when they were admitted to the local hospital for treatment. One former plant employee told the *Herald* that "when some of these guys come in they are supposed to push a plastic card with their name and social security number and work number into a slot. A lot of them were so drunk they couldn't even get the cards in the slot. I'm talking about legless. Now I don't mind if they were pushing a broom or something but these guys had access to the reactor room."[33]

Pilgrim was not the only plant with drug and alcohol problems. At the Trojan nuclear plant in Oregon, fourteen persons, including eleven plant security guards, were arrested on charges of selling various illegal drugs. Officials confiscated 2.2 pounds of cocaine valued at $350,000.[34] (Alcohol and drug abuse have also been problems among military personnel with access to nuclear weapons.)[35]

Another example of *people-related problems* in the nuclear industry is the case of the South Texas Nuclear Project, a twin-reactor plant being built by a consortium of utilities headed by the Houston Lighting and Power Company. At the South Texas Nuclear Project, *quality assurance* and *quality control* programs (*QA/QC*), a vital component of the blueprint for nuclear safety, were subverted to speed construction. An NRC special investigation, completed in April 1980, revealed that QA/QC inspectors, who were supposed to assure that all work complied with NRC safety standards, were harassed, threatened, and intimidated by construction foremen.[36] One inspector testified that a foreman threatened to hit him with a shovel, and said "he would be waiting . . . in the parking lot with a .357 magnum."[37]

The threats persuaded some inspectors to approve defective work, to *"pencil-whip"* reports rather than to conduct inspections. Faulty concrete-pouring techniques were used, leaving holes, or *voids*, within the walls of the reactor containment vessel.[38] The NRC's investigation also found defective welds.[39]

The most serious *people-related problem* in the nuclear power program since the Three Mile Island accident is the failure of many nuclear developers to modify their attitudes about nuclear safety in response to the accident. The nuclear mindset is deeply entrenched, and the Kemeny Commission was concerned that nuclear developers might not make the changes in attitudes that it recommended: "No amount of *technical 'fixes'* will cure this underlying problem. . . . As long as proposed improvements are carried out in a *'business as usual'* atmosphere, the fundamental changes necessitated by the accident at Three Mile Island cannot be realized."[40]

The Rogovin Report, which was issued nine months after the accident, criticized the NRC's *"business as usual* approach,"[41] warning that nuclear developers did not seem to be learning the lessons of Three Mile Island: "Just as the last major reactor accident, the Brown's Ferry fire, slipped beneath the surface of the sea of daily concerns 4 years ago, so can Three Mile Island join it in the coming years. It will take dogged perseverance in the nuclear industry and in the Government to truly learn the lessons of TMI. We are not reassured by what we see so far."[42]

In September 1980, a year and a half after the accident, the Nuclear Safety Oversight Committee, which had been chartered by President Carter to advise him on progress being made in implementing the recommendations of the Kemeny Commission, criticized the NRC's *"'business as usual'* mindset*."* The committee said that there were still "fundamental problems of leadership and attitude within both the industry and the Nuclear Regulatory Commission."[43]

Just as the NRC learned the "too favorable lesson" from WASH-1400, some nuclear developers seem bent on learning the "too favorable lesson" from Three Mile Island. Writing in the June, 1981 issue of *Scientific American*, Harold M. Agnew, president of General Atomic Company and former director of Los Alamos, provided a good example:

> The experience at Three Mile Island demonstrated to the satisfaction of *technically qualified* people that present-day water-cooled nuclear reactors offer *no significant threat* to the health and safety of the general public. . . .[44]

Agnew went on to say that Three Mile Island also showed that such accidents could involve high financial risks, and in "the *extreme case* an accident such as the one at Three Mile Island can threaten the financial survival of the operating utility."[45] (General Public Utilities, the owner of Metropolitan Edison, has been teetering on the edge of bankruptcy ever since the accident; estimates of the cost of cleaning up and repairing the plant exceed a billion dollars, and will probably continue to climb.)

Agnew's statement is interesting for two reasons. First, his assertion that TMI showed that nuclear accidents pose *no significant threat* to public safety is not supported by either the Kemeny Report or the Rogovin Report. Indeed, this assertion is a product of precisely the attitude that the Kemeny Commission criticized: the belief that nuclear power plants are *sufficiently safe*.

Second, Agnew's use of the phrase *technically qualified* persons illustrates one of the ways the nuclear mindset defends itself. The claim that *technically qualified* people agree that reactors pose *no significant threat* provides an automatic justification for ignoring anyone who interprets the *facts* differently. Those who do not share Agnew's mindset are—by implication—*unqualified*. Thus the mindset perpetuates itself, and business proceeds as usual.

PART FIVE

Leaks, Spills, and Promises

The development of disposal of liquid [radioactive] waste in salt, I personally believe, can be worked out in the next two or three years. It is not so much whether it can be done. It is just how to do it. In the case of salt, I honestly believe that we could work this out in the next three years.

E. G. Struxness
Oak Ridge National Laboratory
1959[1]

Except for the modest effort on salt, the geological aspect of the HLW [high-level radioactive waste] repository problem had largely been neglected by our generation until a year or so ago. It will not be solved without a strong commitment of money and manpower, lasting beyond 1985.

Ad Hoc Panel of Earth Scientists
Report to the Environmental
Protection Agency
1978[2]

One characteristic of the nuclear mindset is the belief that the problems of nuclear technology will yield to technical solutions that are both practical and affordable. In keeping with this belief, nuclear developers assumed that they would easily be able to find an *acceptable* solution to the problem of *nuclear waste management*. The history of radioactive waste policy illustrates the tenacity with which nuclear developers have clung to this belief in the face of evidence that contradicts it.

Since the beginning of the search for a method for the permanent disposal of highly radioactive waste, nuclear developers have systematically failed to distinguish between hopes for solutions and demonstrations of workable technologies. Even as tanks leaked, demonstration facilities failed, and timelines slipped, nuclear developers somehow continued to believe that the waste problem could be easily solved. This optimism was so strongly held that many nuclear developers viewed radioactive waste management primarily as a public-relations problem.

The extraordinary nature of the materials in radioactive waste makes the task of managing it a complex and difficult one. *High-level* nuclear waste—such as the material produced by fissioning nuclear fuel in a reactor—is extremely radioactive and long-lived. It must be kept out of the environment for thousands of years. Such long isolation requires almost perfect containment during shipping, processing, and temporary and long-term storage.

Low-level nuclear waste—which consists of equipment, clothing, filters, tools, and machinery that have been contaminated with radioactive material—is produced in enormous volume. It is much less radioactive than high-level waste, though *hot spots*—bits of very radioactive material—may be mixed in with the less radioactive bulk.

Nuclear developers have proposed a number of methods for disposing of high-level nuclear waste, ranging from injecting it into deep wells, to implanting it in the sea floor, to depositing it in the polar ice caps, to shooting it into outer space. Technical problems and uncertainties plague all these methods, and many of them have been abandoned. The most popular disposal concept is the idea of burying the waste in underground *repositories* deep within *stable* geologic formations. Geologic uncertainties limit the ability of scientists to predict what will happen to a repository over a long period. The waste-disposal problem remains unsolved.

CHAPTER TWELVE

A Nonexistent Problem

The search for a method of disposing of high-level nuclear waste began in an atmosphere of optimism and ignorance. In the wartime rush of the Manhattan Project, scientists—not surprisingly—did not have time to worry about the long-term fate of the high-level waste produced at Hanford, Washington. The waste was stored in carbon-steel tanks, and the search for a long-term plan was deferred.

After the war, the AEC continued to rely on tank storage as its nuclear weapons complex grew. AEC officials developed a simple maxim to describe their theory for controlling radioactive waste: for high-level wastes, *concentrate and contain*; for low-level wastes, *dilute and disperse*.[3] Confining high-level waste in tanks would permit the AEC to develop a method for permanently disposing of the troublesome substance. According to policy, high-level waste, stored in carbon-steel tanks, could remain in *interim storage* until a method of *permanent disposal* was found; low-level waste could be buried in trenches (the current policy) or dumped into the ocean.

More than three decades later, the search for a way to dispose of high-level waste continues. In the absence of a method for *permanent* disposal, *interim* solutions have become increasingly permanent. High-level waste from the weapons program continues to be stored in tanks, many of which are leaking. The nuclear power industry's *spent fuel rods*—also a form of high-level waste—have piled up in temporary *spent fuel pools* at the nation's reactor sites. Since there is no place to take the waste, many utilities have expanded their spent fuel pools to make room for the growing *waste inventory*. Despite the poor waste-management record and a long list of past mistakes, many nuclear developers remain reluctant to admit that there have been—and still are—serious problems in the waste-management area.

In the late 1940s, nuclear developers did not believe that nuclear waste would ever present a problem, and the newly created AEC was too busy building bombs for the nation's atomic arsenal to worry much about waste disposal. Also, it seemed that waste disposal could be achieved with relative ease when compared with reactor design. Besides, the tanks at Hanford

Projected Need for Waste Repositories

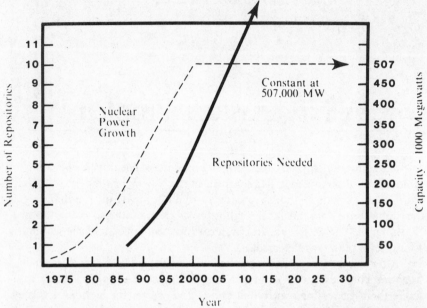

GRAPHIC BY JOHN HEYMANN

FIGURE 11 How the size of the nuclear power industry affects the number of radioactive waste repositories needed. The dotted line shows growth of the nuclear power industry, according to optimistic industry growth projections through the year 2000. After the year 2000, growth is assumed to stop. The solid line shows the number of high-level nuclear waste repositories of the size and type currently envisioned by the DOE that would be needed to accommodate the waste produced. Even with the assumption that industry growth would stop at 507 gigawatts, a new repository would be needed every two or three years.

Source: Adapted from State of California, Nuclear Fuel Cycle Committee, California Energy Resources Conservation and Development Commission, *Status of Nuclear Fuel Reprocessing, Spent Fuel Storage and High-Level Waste Disposal,* draft report, 11 January 1978, p. 218.

were in place and were believed to be quite adequate as an interim solution. As Abel Wolman—a waste-disposal specialist from Johns Hopkins University who presented the opening remarks at the hearing of the Joint Committee on Atomic Energy on waste disposal in 1959—told Congress, "the problem of radioactive waste was considered to be *nonexistent*" during the first decade or so of atomic development.[4]

Interest in radioactive waste picked up only slightly after the Atomic Energy Act of 1954, which opened up the nuclear field to private industry.

As Wolman testified, "It took a good deal of time before many people interested in the total nuclear fission field were willing to acknowledge the existence of an expanding difficulty in radioactive waste."[5]

The failure to recognize the existence of the waste problem was due to the optimistic attitude that confused ideas still in the laboratory stage with practical technologies. Wolman testified in 1959 that there had been "some tendency" to "announce treatment processes . . . which have a *high imaginative content* but *low ingredients of reality*."[6]

A second reason for the lack of attention to waste disposal was, as a 1957 AEC report pointed out, that "industry has felt that waste disposal was not a particularly attractive *profit-making undertaking*,"[7] and did not want to get directly involved. The AEC was left with the responsibility for developing and operating facilities for safely disposing of radioactive waste, both from its own military program and from commercial industry.

The fact that the government had assumed responsibility for waste management had a profound impact on the attitudes of commercial nuclear power developers. Wolman said that since the AEC is responsible for taking care of the radioactive waste produced by industry, "it disappears from the account book of the power developer." As a result, Wolman said, "industrialists interested in nuclear fission power . . . consider the waste problem to be quite unimportant. . . . It is unimportant to them because they are not responsible for its management and hence its cost."[8] (Present policy calls for the utilities to pay the federal government a *one-time fee* in exchange for which the government will assume the burden of managing the wastes indefinitely.)

A history of U.S. waste-management policy published in 1978 for the NRC by political scientist Daniel S. Metlay concluded that the AEC's attitude toward waste management was marked by "periods of unconcern interspersed with moments of intense interest. Lacking the *sex appeal* of reactor development and the pork barrel quality of other segments of the fuel cycle, waste management became, organizationally and operationally, a residual category."[9]

Metlay reported that the AEC commissioners were not interested in waste. Even in the 1970s, Commissioner Dixie Lee Ray, for example, reportedly would "turn up her nose" when the subject came up.[10] In the AEC ranks, the personnel figured out that "grand careers were made in reactor development where the organization's resources were committed, not in waste disposal."[11]

The AEC made public relations a guiding force in shaping waste-management policy. The AEC's 1957 report on the status of waste-disposal programs, for example, said that "public relations" in the waste-management area will be an "important facet to be considered as the industry expands." The report suggested that "as the public becomes more aware of the environmental and health problems" it will "prove advantageous" for the AEC and other agencies to carry out *"positive educational programs"* on what is being done to *"control"* waste problems.[12]

A 1954 report on radioactive waste disposal in the ocean illustrates this concern with public relations. The report, a set of proposed guidelines for ocean dumping of low-level waste, devotes an entire section to the discussion of the *"public relations aspects"* of sea disposal, stressing the need to avoid *"undue public alarm."*

> Any recommendations concerning the disposal of radioactive wastes in the ocean shall not only provide *adequate safety* but also be such that they will minimize (or if possible eliminate) the possibilities of *undue public alarm*. *Unfavorable situations* might arise if a package of radioactive material were found on the shore or recovered in a fisherman's net or by a trawler or dragger.[13]

The report also cautioned that public relations at the international level must not be forgotten: "Last but not least, it is possible that waste disposal practices may be responsible for international incidents leading to formal protests being lodged between nations."[14]

Between 1945 and 1969, the U.S. dropped tens of thousands of canisters of low-level waste into the ocean.[15] (In 1972, ocean disposal of nuclear waste was banned by the U.S., though Great Britain continues to dump low-level waste into the sea. Currently, low-level waste in the U.S. is buried in shallow trenches.) U.S. ocean disposal was marked by shocking practices and managerial incompetence.

In 1980, the *Boston Globe* did an exposé on the dumping practices of Crossroads Marine Disposal, operated by George Perry and John Santangelo:

> *Crossroads* was named after an atomic bomb experiment both to describe and to disguise its work.
>
> "We couldn't use the word *'nuclear'* in those days," Perry said. "They [the Atomic Energy Commission] didn't want us to. It was the fright of it they wanted to avoid." . . .
>
> For the most part, Santangelo and Perry were on their own. "It was pretty much of a free-for-all in those days with what you did with your *stuff*," Santangelo recalls.
>
> Not until 1960, Perry says, did any government agency monitor Crossroads' operations. Then the Army Corps of Engineers started sending an inspector out with each load. . . . Perry says that while the inspector's job was "to make sure we hit the mark," the government man "was usually drunk by the time we got there. We made sure of that." Veteran employees at the Corps office in Waltham, in turn, remember Perry as "a real cowboy." . . .
>
> While the Atomic Energy Commission licensed Crossroads to dump only at a specific site, Perry says puckishly that he can't be certain it all

landed there. Navigational tools were less sophisticated then, he says, adding: "Sometimes I couldn't find the dumping ground." . . .

Later, Perry says, to replace the earlier practice of shooting holes in drums to make them sink, he invented a valve that would flood a container with seawater as it went down, then close automatically when the cargo bumped bottom. . . .

Perry says he invented the only recorded number about how much radioactivity Crossroads dumped off Cape Ann.

He did this in 1959 when the Atomic Energy Commission, on the eve of the hearings before the Joint Committee on Atomic Energy that eventually brought a halt to ocean dumping, pressed him to declare how much he had dumped.

Perry said his early records were lost in a 1956 fire in his T-Wharf office, and at best he could only guess. "George, we need an *answer*," he says he was told. He provided the one that was offered as testimony: 4,008 containers containing 2,400 curies of radioactivity.

"I didn't even know what a curie was," Perry guffawed. . . .[16]

CHAPTER THIRTEEN

Hanford: The Interim Reality

The record of nuclear waste management at the U.S. government's Hanford Reservation does not inspire confidence in the ability of nuclear developers to control radioactive waste. Tank storage—originally billed as an interim solution—has become a permanent problem. Hundreds of thousands of gallons of high-level waste have leaked from Hanford's underground storage tanks. The leaks are hard to detect, since measurements of the liquid level within the tanks are uncertain. Nevertheless, few nuclear developers have questioned the assumption that they can manage nuclear waste safely. The waste managers at Hanford have proved incapable of containing the waste. When the tanks leaked, however, they devoted much attention to containing information on their waste-management failures.

Hanford is a 570-square-mile site built alongside the Columbia River in a remote desert in southern Washington, set up by the Manhattan Project to produce plutonium for the first atomic bombs. Over the following two decades, Hanford expanded into a giant industrial complex. The AEC constructed nine plutonium production reactors and a host of reprocessing plants and other kinds of chemical processing plants. So great was the AEC's sense of Cold War urgency about producing large nuclear weapons stockpiles that the agency constructed the first eight reactors with once-through, single-loop cooling systems: water from the Columbia River was pumped directly over the reactor cores and discharged back into the river. (The last of the once-through reactors was shut down in 1971.) Spent fuel from the plutonium production reactors at Hanford was reprocessed, the weapons material was shipped to weapons fabrication facilities, and the high-level nuclear wastes were pumped into underground tanks as an *interim solution* to the waste problem.

The tanks at Hanford, which range in size from 55,000 gallons to 1,000,000 gallons, were constructed of carbon steel and concrete (see plate F). The first 149 tanks, built between 1943 and 1964, are *single-shell* tanks, consisting of a carbon-steel liner surrounded by reinforced concrete and covered by a dome. Since 1968, twenty *double-shell* tanks, with two separate carbon-steel liners, have been added.[1] The tanks are grouped into clus-

PLATE F Million-gallon radioactive waste storage tanks under construction in Hanford, Washington. The tanks shown are designed to store high-level liquid wastes from the weapons program. Constructed in a cavity carved into the soil, the tanks will be buried 15 feet underground after the excavation is backfilled. In the background is the PUREX plant, a reprocessing plant for producing plutonium and uranium for nuclear weapons.
Source: U.S. Department of Energy.

ters called *tank farms*, and they are woven together by an extensive network of transfer lines for pumping waste from one tank to another.

The waste managers at Hanford did not expect the tanks to begin leaking for a number of decades. As Herbert Parker, manager of the Hanford Laboratories, told the Joint Committee on Atomic Energy in 1959, "these wastes will have to remain isolated from the environment" for "a longer time than any operation heretofore contemplated by man."[2] Parker testified that in normal industrial applications, tanks like the ones at Hanford would be expected to last for 100 to 200 years.[3] But he added that "we have to report that the life of the tanks is not yet known. None has as yet failed and the life is estimated to be at least several decades." He told the committee that "until we create a better way," the waste will have to remain isolated in the tanks, though not necessarily in the same set of tanks. "In other words," he said, "if the tanks we have turn out to have a life of 50 years, it will be *very simple* to be prepared at the right time with an alternate set of tanks and pump the liquids into new tanks. We have extensively moved the liquid from one tank to another and we are persuaded that we can do this with *perfect safety*."[4]

At the same hearings, the question of whether any of the tanks had leaked came up several times. When pressed about this, Parker replied:

> We have had what might be described as *suspicious occurrences* in these tanks. With a tank as large as this it is difficult to create suitable devices, since there is no access to the device, to measure the liquid level with precision. There have been times when there have been *apparent oscillations* in these liquid levels and *substantial investigations* as to whether the material has in fact leaked have been made. If we answer the question in that framework, we have *never detected* a leak from any of these tanks, so we are in turn persuaded that none has ever leaked.[5]

Parker's statement provides a striking illustration of the workings of the nuclear mindset in the presence of uncertain data. The reason Parker and his colleagues were "persuaded" that none of the tanks had ever leaked—despite the "suspicious occurrences" and "apparent oscillations"—was because they wanted to believe that no tanks had leaked. Because of that desire to believe, they interpreted the data on liquid-level oscillations and their uncertainty about the data to mean that there had been no leaks. Later studies concluded that a first leak had occurred in 1956, and a second happened in 1958.[6] Parker and his colleagues did not *detect* these leaks initially—despite warnings from their instruments—because of their mindset.

The waste managers at Hanford have denied the public, and the scientific community access to reports and data on tank leaks. In 1953, the U.S. Geological Survey completed a report on the geology of the Hanford site which noted that the tanks were a "potential hazard" and pointed out that their "true structural life . . . is not entirely known."[7] The report, one of the first to question the wisdom of tank storage, was classified by the AEC. It was declassified in 1960, but was not published in the open literature until 1972.[8]

In 1968, the U.S. General Accounting Office completed a report that was highly critical of waste-management practices at Hanford, saying that some tanks were being subjected to stress "well beyond accepted design limits," and charging that tank farm operators had continued to use tanks with known weaknesses when there were no others available.[9] The report was kept secret until December, 1970, when political pressure forced its release. The AEC claimed that the report had been classified not to avoid embarrassment but to protect information that would make it possible to calculate plutonium production rates. After the report was declassified, AEC officials termed keeping it secret *"overly cautious."*[10]

Meanwhile, leakage problems continued to grow more serious. In the early 1970s, leaks began appearing more frequently, and their sizes increased.[11] In June, 1973, Hanford officials detected a leak in tank 106-T— the largest leak so far.[12] About 115,000 gallons *escaped engineered control*, raising serious questions about the AEC's ability to detect leaks in a *timely fashion*. An estimated fifty-one days of steady leaking occurred before the 106-T spill was discovered.[13] The reason: the waste managers were too busy with what one called "the press of other duties" to read and interpret the data that could have told them the tank was leaking.[14]

The 106-T accident raised some unsettling questions. As Robert Gillette wrote in *Science:* "Is the AEC really prepared to manage thousands of pounds of wastes that civilian nuclear power plants will be generating in the years ahead? And how, exactly, could it lose the equivalent of a railroad tank car full of radioactive liquid hot enough to boil itself for years on end and knock a Geiger counter off scale at a hundred paces?"[15]

In response to the political reaction to the 106-T leak, Hanford adopted more stringent procedures for tank-farm monitoring. An Energy Research and Development Administration report in 1975 said that due to "improved leak detection systems," a leak as large as 106-T "is not *expected* to occur again under *normal operations*."[16] But leak detection is an inexact procedure, and great uncertainties call into question the accuracy of the waste managers' optimistic claims.

Historically, two methods have been used to detect leaks at Hanford. The first is the use of probes, which are supposed to locate the surface of the liquid and register the liquid level. Once a loss of liquid is detected, a judgment must be made as to whether the change is due to tank leakage or some other cause, such as evaporation.[17]

The second method of leak detection is the use of *dry wells*, steel-lined holes drilled in the soil around the tanks. Periodically, radiation meters are inserted into the dry wells to measure changes in the radiation levels in the soil surrounding the tanks.[18]

The two leak-detection methods are considered to be mutually supportive. If a probe detects a loss of liquid in a tank, the reasoning goes, radioactive liquid seeping into the soil will also be detected by the dry wells. If an increase in radiation levels is detected in the dry wells, then (if the cause was a tank leak) the probe should indicate a liquid loss.[19]

This approach sounds straightforward, but matters are not so simple. The main difficulty is that the waste no longer consists of liquids only, but of a multilayered mixture of liquids and solids, making it difficult to measure liquid levels. The probes are unable to measure the level of liquid trapped in the *sludge* and *salt cake* at the bottom of a tank.[20] And though the probes can measure the level of liquid that is resting on top of the salt cake, the accuracy is *perturbed* by stalactites, crust,[21] and chunks of floating salt cake. As a result, measurements of liquid levels are quite uncertain.

Dry-well measurements also have their limitations. First, heightened radiation levels in the soil near a tank could be caused by the migration of an old spill rather than the presence of new leakage. Second, it is expensive to install and monitor dry wells, and there has been a debate among Hanford personnel about how many are needed. As Stephen Stalos, manager of *Tank Farm Surveillance Analysis* at Hanford, said in 1978, "Tanks which have the good sense to leak in the immediate vicinity of dry wells will indeed be quickly detected. But how many tanks have such good sense? . . . [A] leak of hundreds of thousands of gallons of nuclear waste may never be detected in dry wells, depending upon their position relative to the leak. . . ."[22] Despite Stalos's call for additional dry-well monitoring, the DOE and the current Hanford contractor, Rockwell Hanford Operations, decided to reduce dry-well monitoring in 1978. Stalos resigned in protest, charging that the cutback was part of a politically motivated coverup being carried out by Rockwell "at the request of the Richland Office of the Department of Energy."[23] According to Stalos, the reduction in monitoring—as well as a decision not to announce three leaks that had been detected—were part of this coverup. Finally, Stalos charged that a Hanford official had told him it was Hanford policy that there would be "*no more leaks*."[24]

Stalos's charges prompted an investigation by the DOE's inspector general. His report, issued in 1980, sharply criticized waste-management practices at Hanford, saying that certain policies and practices were in need of "wholesale overhaul."[25] (In its comments on the inspector general's report, DOE's Richland office suggested that the term "*wholesale overhaul*" be replaced with the word "*revision*."[26])

One of the areas the inspector general criticized was the Hanford system for classifying tanks. Currently, there are four categories of single-shell tanks: Active tanks, Questionable Integrity tanks, Confirmed Leakers, and Inactive tanks. These classifications date back to 1973 and are defined as follows:

Active tanks include all tanks that are "in regular current use and that have no known flaws." Active tanks can be used for the storage of unsolidified liquid waste.[27]

Questionable Integrity tanks are tanks where either an unexplained liquid-level drop has been detected, or heightened dry-well radiation levels have been found, but not both.[28]

Confirmed Leakers are tanks where both liquid-level decreases and heightened dry-well radiation readings have been detected and are believed to be caused by tank leakage. Sometimes, "when there is no doubt that a tank is leaking," it will be classified as a Confirmed Leaker on the basis of liquid losses alone.[29]

Inactive tanks include all tanks not classified as of Questionable Integrity or as Confirmed Leakers which have been retired from "active service." This means tanks not used for storing unsolidified liquid waste. Inactive tanks continue to be used to store salt cake.[30]

Tanks classified as either of Questionable Integrity or as Confirmed Leakers are referred to as *unsound*, while Active tanks and Inactive tanks are considered to be *sound*. Of the 149 single-shell tanks, twenty-five are Active, thirty-four are judged to be of Questionable Integrity, twenty-four are Confirmed Leakers, and sixty-six are Inactive.[31]

There is no *operational difference* between Questionable Integrity tanks and Confirmed Leakers—both are considered to be unsound. But until a recent change in policy, there was one crucial difference in the way the two categories were handled: only when a tank was classified as a Confirmed Leaker were the news media notified.[32] This policy helped keep publicity about tank leaks to a minimum. Prior to the big leak from tank 106-T in 1973, no public announcement at all was made when it was concluded that a tank was leaking.[33] Following the 106-T leak, it was decided that press releases would be issued when a tank was classified as a Confirmed Leaker, but that the media would not be informed when tanks were classified as of Questionable Integrity. The inspector general found that this policy was never put in writing and was unable to determine exactly how it was adopted.[34]

Since the waste managers have a hard time obtaining the data necessary to declare a tank a Confirmed Leaker, the policy of announcing only *confirmed leaks* greatly reduced adverse publicity. During the years 1973 through 1975, there were six public announcements of tank leaks.[35] In 1977, a Hanford official was able to tell Congress that there had been "*no leaks*" for two years.[36] And between July 1975 and January 1980, no tanks were classified as Confirmed Leakers. As a result, there were no press releases on tank leaks for four and one-half years.[37]

In January 1980, Rockwell Hanford Operations released its most recent "*tank status reclassification report*," which concluded that four Questionable Integrity tanks should be *reclassified* as Confirmed Leakers.[38] At the news conference announcing the reclassification, Paul Fritch, assistant general manager for Rockwell Hanford, was careful to "emphasize that this does not mean that there are now four *new leaks* at Hanford...," saying the "leaks we are confirming today occurred more than nine years ago."[39]

The use of the word *new* in this context is misleading. Since the tanks had begun leaking some time before, Fritch's statement that the four leaks were not "new leaks" may have been literally true. But it also illustrates a subtlety in the way waste managers view tank leaks: at no time during its history is a leak considered to be a *new leak*. When it is initially detected, it is not a new leak because it is not yet *confirmed*—it is only a *possible leak*. Once it is finally confirmed, it is not a new leak because it occurred years before. Responsibility for the leak is always shuffled into the past or deferred into the future.

The 1980 tank reclassification study was the subject of a sharp disagreement between the inspector general's office and Rockwell Hanford personnel. The dispute was detailed in the inspector general's report in a section titled "A *Disquieting Episode* Relating to the Reclassification Study." A draft of the reclassification study was completed several months before the inspector general began his investigation. Yet none of the Hanford officials mentioned the reclassification study to the investigator from the inspector general's office. Later, the investigator learned that a meeting of Rockwell waste-management officials had been held during which it was decided that the existence of the draft study not be mentioned to the investigator, though the study would be provided if he requested a copy. The inspector general called "the failure of these officials to have volunteered the report ... a matter of disappointment and concern."[40]

One of the officials, John Deichman, program director for waste management, countered that the report was not mentioned because it was not "far enough along." Rockwell, Deichman explained, did not want to burden the inspector general with extraneous information. "The inspector general expected to be told by everyone of the existence of everything," he said, adding that the "inspector general made a big deal out of something that wasn't all that large."[41]

The inspector general was unable to determine whether it was true, as Stalos charged, that there was a policy to cover up tank leaks. Stalos and the people Stalos had talked with gave different accounts of their conversations. There was no way to ascertain who was lying. The inspector general was unable "to arrive at a judgment." But he concluded that, because of the procedures followed at Hanford, a coverup was unnecessary—it was built into the system.

The word *"cover-up"* evokes pictures of people devising strategies and tactics aimed at concealing things which ought not to be concealed. But in the case of Hanford, had there been any officials desiring to minimize publicity about tank leaks, they would have had no real need to engage in conduct which might be considered questionable. This is because Hanford's existing waste management policies and practices have themselves sufficed to keep publicity about possible tank leaks to a minimum.[42]

Thus, a coverup would take place in the normal course of business.

At Hanford, part of the normal course of business is the redefinition of familiar words, by which problems are defined out of existence. One little-publicized but grotesque example is the case of the BC cribs.

A *crib* is an underground structure designed for discharging liquid wastes directly into the ground. It consists of a ditch about 20 feet deep and up to 1,400 feet long, filled with rock and covered with an impermeable membrane and soil. A pipe running the length of the crib releases the waste uniformly along the crib length, where it percolates into the soil. One hundred seventy-seven cribs have been constructed at Hanford.[43]

Cribs generally are used for the disposal of low-level and *intermediate-level liquids*—the latter being liquid waste with somewhat higher levels of radioactivity than low-level liquids. Between 1953 and 1957, however, the AEC intentionally discharged high-level liquid waste directly into the BC cribs. The AEC took this extraordinary step because there was not enough room in the tanks to handle the quantities of waste Hanford was generating.[44] Plutonium production took priority over environmental protection, so the waste was simply dumped into cribs. No estimate is available of the total amount of radioactive waste that was deliberately dumped,[45] but the cribs and trenches in the BC area received the greatest amount of radioactivity discharged at any one site on the Hanford Reservation. Though the exact nature of the wastes dumped in the BC area is not available, it is known that an estimated 900,000 curies of beta-emitting fission products were disposed of at that site.[46]

Thomas Bauman, director of DOE's public affairs office at Hanford, objects to the use of the term *high-level* when referring to the waste dumped into the BC cribs. "The wastes disposed of were *intermediate level* wastes," he explained, "which through a *redefinition of terms* several years later fell into the redefined 'high-level' category."[47] Bauman's verbal maneuvering obscures the situation; no matter what you call them, the wastes in the BC cribs are extremely radioactive, containing high concentrations of long-lived fission products.

Interviewed in 1980, Frank Standerfer, assistant manager for technical operations at the DOE's Richland Operations Office, said that the practice of dumping fission products into cribs had been discontinued and that high-level waste dumping would not be done today. But Standerfer defended the decision to dump the waste during the 1950s, saying that it is "not viewed as all that undesirable" because the waste "will stay where it is until it decays." Standerfer claimed that monitoring of the BC crib sites showed that the situation is "very *stable*."[48]

Standerfer did not mention that the BC crib site had been *breached* and that radionuclides had been spread over a sizable area. In 1960 animal burrows were discovered in residue salts at the bottom of the BC trenches.[49] Animals—presumably at least one badger—had exposed a salt layer.[50] Rabbits subsequently used the sites as a salt lick, spreading radioactive feces and urine over 2,100 acres of sagebrush-covered desert.[51] Radioactive coyote feces were also detected.[52] A four-square-mile area was cordoned off and designated the BC *Controlled Area*.[53]

Standerfer's opinion that the BC crib situation is "stable" was echoed in ERDA's environmental impact statement on waste management at Hanford: "The bulk of the radioactivity remaining is *fixed* in *rabbit droppings* scattered over approximately 4 square miles of ground surface."[54]

The use of the words *fixed* and *stable* in this context is outlandish, since rabbit droppings are hardly formed of a material that will remain unchanged for centuries. The words *stable* and *fixed* imply that the radioactive materials, spread over 2,100 acres of desert terrain, will not move along biological pathways. But some of the radioactive material in the animal droppings has already been taken up by sagebrush and other vegetation. This presents a situation which is very *un*stable. Elsewhere in the environmental impact statement, there is a discussion of the possibility that a range fire in the BC Controlled Area could "serve as a *transport mechanism*." If a fire occurred, burning would release radioactive material in the vegetation into the air. Subsequent wind erosion would transport additional material from the topsoil. Range fires on the Hanford Reservation are not uncommon. An average of twelve fires occur every year, and these fires cover areas varying in size from less than 1 acre to 32,000 acres.[55]

The future of the high-level waste in the tanks at Hanford remains uncertain. Hanford officials admit that more tank leaks are bound to occur. As Deichman said, "All of the tanks are going to leak—it's just a matter of time. . . . With the technology we have today, there's no excuse for having liquid wastes. But we are locked into a system, and we've got to get out of it."[56]

In the search for a permanent solution, the waste managers at Hanford are investigating a number of alternative methods. One proposal is to *reinforce* the existing tanks and leave the waste where it is. Another plan calls for removing the waste from the tanks, solidifying it, and disposing of it in a geologic repository, possibly constructed in the basalt underlying the Hanford Reservation. No decision has been made about what method to use for *ultimate disposal*.

In 1977, ERDA outlined twenty-eight alternatives for long-term management of the Hanford waste. Though the technology required to implement any of these plans has not been fully developed,[57] ERDA made risk assessments and cost estimates using *preliminary design information*. The cheapest plan, leaving the waste in existing tanks, would cost an estimated $500 million (1976 dollars); while the most expensive alternative would cost an estimated $6 billion.[58]

If a decision is made to leave the waste in the existing tanks, the tanks would first be *"hardened"* so they would remain *reliable for centuries.*[59] Hardening would involve *"additional engineered containment,"* which could include "filling of the void between the dome and the waste with sand or other suitable material" to help prevent tank dome collapse, reinforcing tank walls with additional concrete, placing a "confinement barrier of concrete over the tops of the tanks," piling crushed stone and cobble over the tanks to prevent erosion, and installing "an underground radiation surveillance system" for ongoing monitoring.[60]

Deichman said that the *leave-or-retrieve* decision about the wastes stored in the single-shell tanks will be made about 1985. He stressed that there is no hurry: "The tanks are *probably adequate* for tens of years." And we "may want to wait until the year 2300 to take the waste out of the tanks," he added, since the radioactive decay of the waste would make it easier to work with at that time.

Deichman said that initial drafts of a report on the leave-or-retrieve decision will be available to the public within the next few years. Before the document is "cleared for public release," he said, it will be carefully *"word-engineered."*[61]

CHAPTER FOURTEEN

The Permanent Nonsolution

The idea of geologic disposal of high-level waste—considered the most promising long-term disposal method—has frequently been the basis of optimistic statements, programs, and timelines. Since the late 1950s, the AEC and its successors have repeatedly declared that a functional repository would be operational within a few years. But insufficient understanding of geologic processes has made long-term geologic predictions very uncertain. The reassuring statements proved to be exaggerations, and the opening date of the first repository has receded into the future as complications have arisen. According to the latest federal timeline, established in 1980, nearly thirty years after the geologic concept was first proposed in 1951,[1] a permanent waste repository will be operational by the mid-1990s.[2]

Geologic disposal would work in the following manner—at least according to theory: a carefully selected rock formation, geologically *stable* and free of groundwater, would keep the waste isolated from the environment for thousands of years. The type of rock and the repository site would be chosen so as to minimize the likelihood that the waste would be transported into the biosphere—either by groundwater, or by *human intrusion*, or by some other *pathway*—before its radioactivity has decayed. The federal government's current plan for disposal of high-level and TRU waste depends on successfully putting this theory into practice. (*TRU waste* is a special category of waste contaminated with transuranic elements—elements like plutonium that are heavier than uranium and whose radioactivity is especially long-lived.)

In 1955, the National Academy of Sciences established a multidisciplinary group to examine various options for disposal of radioactive waste on land.[3] Two years later, the academy suggested that salt might be a good geologic medium in which to bury waste. Salt is water-soluble, and the presence of a salt deposit in an area indicates that the area is free of groundwater. Salt also flows such that boreholes and tunnels cut for waste emplacement would seal back up. Great enthusiasm for salt disposal quickly developed. E. G. Struxness of the Oak Ridge National Laboratory told

Congress in 1959 that he believed that "we could work this out in the next three years."[4] But other scientists were not convinced. R. L. Nace, associate chief of the Water Resources Division of the U.S. Geological Survey, told Congress at the same hearing:

> The big questions are, Can the hazardous materials be immobilized? If they move, how will they affect the usability of water supplies? What paths will they follow? Where will they stop? How long will it take? and, Will they be concentrated or dispersed? Even when we can answer these questions, we will not have really solved any problem. We will have merely described a situation.[5]

Progress on geologic disposal has been beset with problems, and twenty-two years later, most of Nace's "big questions" are still unanswered.

The AEC began testing the salt-disposal idea in 1963 at a site known as *Project Salt Vault*, an experimental repository in an abandoned Kansas salt mine. A great deal of data on the effects of radiation and heat on salt were generated. But the AEC, certain that the waste problem was under control, had lost interest and paid little attention to the research.[6]

A fire at the AEC's weapons plant at Rocky Flats, Colorado in 1969 forced the agency to reopen its investigations of long-term disposal. The fire produced a tremendous quantity of low-level plutonium-contaminated debris, or, in today's jargon, TRU waste, though that term was not yet in use. This waste was shipped to the AEC's Idaho Falls Reactor Testing Station for burial. Idaho Senator Frank Church was outraged by this action and extracted a commitment from AEC Chairman Glenn Seaborg that all of this waste would be removed from the state by the end of the 1970s.[7] In order to keep this promise, which was a matter of political importance, a repository had to be built at once.

The public release of a National Academy of Sciences report critical of AEC waste-management policy made the search for a repository site even more urgent. The AEC had suppressed the report after it was completed in 1966, but under pressure from the U.S. Senate, the report was released in 1970.[8] The report warned that considerations of long-range safety had in some instances been subordinated to regard for economy of operation. The Rocky Flats fire and this report prompted the AEC to propose transforming Project Salt Vault into a permanent repository. That the AEC expected the proposed waste repository to receive a *"favorable reception"* from the local citizens[9] also contributed to this politically motivated decision.

On June 12, 1970, the AEC decided to establish a *"first-of-a-kind demonstration facility"*[10] for the permanent disposal of solidified high-level waste and TRU waste in Lyons, Kansas, site of Project Salt Vault. The AEC expected to have the facility fully operational by 1976, saying that it would provide "a *timely permanent solution*" to the problem of waste disposal.[11]

Though selection of the Lyons site was primarily based on the AEC's desire to find a quick fix for the waste problem, the agency publicly claimed that the site was chosen for technical reasons. As one AEC *"Fact Sheet"* said, "The AEC knows of no other site that would be more *suitable"* than Lyons.[12] Even so, the agency announced—in its effort to reassure the public—that it planned to conduct "additional geologic and safety studies . . . to *confirm"* that the site was appropriate.[13]

But additional studies did not "confirm" the wisdom of the AEC's selection. Instead they indicated that the AEC's assessment of the Lyons site as suitable was, at best, premature. In its haste to achieve a "timely permanent solution" to the political problems raised by the waste issue, the AEC neglected to consider many important aspects of the geology of the site. Further investigation revealed that the area was inappropriate for siting a waste repository. The site and the surrounding area were riddled with drill holes, from both abandoned and producing oil and gas wells.[14] As William W. Hambleton, director of the Kansas Geologic Survey, said, "The Lyons site is a bit like a piece of Swiss cheese, and the possibility for entrance and circulation of fluids is great."[15]

The geologic evidence at the Lyons site led the Kansas Geologic Survey, the Kansas Geological Society, and the Council of the Kansas Academy of Science to conclude that the site should be abandoned. Hambleton wrote: "There is nothing more important than recognizing a dead horse early and burying it with as little ceremony as possible."[16]

In addition, further investigations raised questions about the general concept of burying nuclear waste in salt.[17] Independent geologists charged that the mathematical models used by the AEC to predict repository behavior were overly simplified.[18] Even within the AEC, some personnel concluded that calculations that had been described as complete and sophisticated were actually "back of the envelope."[19]

Opposition to the proposed repository grew rapidly, bolstered by evidence of sloppy technical work. The AEC's judgment that it would be easy to achieve *public acceptance* of the Lyons site proved to be as unrealistic as its assessment of the Lyons geology. Beaten on both the technical and the political fronts, the AEC abandoned the Lyons project in early 1972. (The AEC and its successors failed to remove the TRU wastes from the 1969 Rocky Flats fire by the end of the 1970s. These wastes are still in storage at Idaho Falls, where they will stay until the government opens a permanent waste repository.)

Since 1972, the federal government has made several false starts at setting up a waste-management program. A plan to construct a *Retrievable Surface Storage Facility (RSSF)* was scrapped. In the search for an underground repository, exploratory drilling is under way at several locations. The most recent plan was developed by the Carter Administration. If all goes according to schedule, a repository site will be selected by 1985 and will begin operation in the mid-1990s.

Current repository design relies on the *multiple-barrier* approach. This strategy holds that the waste should be isolated by a series of *engineered barriers*, each of which would have to fail for the waste to escape. As a first barrier, the waste would be *vitrified,* that is, solidified in a glass or ceramic *waste form*, which would resist corrosion. The glass or ceramic would be encased in a *canister* made of a material impermeable to groundwater and resistant to corrosion. The canisters would be inserted into the repository, and *near-in stages of containment*, such as metal sleeves or packing material, would form the next barrier. The final barrier would be the geologic formation itself, selected on the basis of criteria designed to ensure that the waste would remain undisturbed for long time periods. Though the barriers might eventually fail, the argument goes, each would introduce a time delay. Once the waste had penetrated all the barriers, its radioactivity would have decayed.

But a number of recent reports and studies cast doubt on the ability of earth scientists to accurately predict the fate of radionuclides emplaced in such a geologic repository. A report prepared for the EPA in 1978 by an "Ad Hoc Panel of Earth Scientists," including geologists from Brown, Harvard, Texas A&M, Princeton, and Dartmouth, raised serious questions about the use of glass as a waste form, saying that it "would not be at all surprising to find that the integrity of the glass was lost over time scales of a decade, instead of the millennia that are now computed. . . ."[20] Noting that "the canister could likely be breached within time scales of a decade or less," the panel concluded, "We do not consider the canister to be a significant barrier. . . ."[21]

The panel also concluded that the near-in stages of containment "cannot be relied upon to effect any significant retardation of the release of the HLW [high-level waste]." The panel defined *significant retardation* as "times longer than a decade."[22] As for the geologic formation itself, the panel noted that there were "uncertainties" involved in forecasting the behavior of hypothetical geologic repositories, "due principally to inadequate knowledge of the relevant mechanical, radiochemical, and hydrologic properties of the candidate rock types." The data could be gathered, but it would likely take "from a year or so to a decade or more."[23]

A circular published in 1978 by the U.S. Geological Survey concluded that a "lack of understanding" of many of the interactions between the mined opening of the repository, the waste, the host rock, and any water the rock might contain "contributes considerable uncertainty to evaluations of the risk of geologic disposal of high-level waste." The circular also noted "geo-chemical uncertainties" associated with the use of salt as a disposal medium.[24] Finally, the report pointed out that geologic predictions over long time spans are uncertain: "Earth scientists can indicate which sites have been relatively stable in the geologic past, but they cannot guarantee future stability. Construction of a repository and emplacement of waste will initiate complex processes that cannot, at present, be predicted with certainty."[25]

Further questions about the use of glass as a waste form were raised by experiments conducted at Pennsylvania State University. A good waste form must have a low leachability (a measure of its resistance to the removal by water of radionuclides in the waste). Tests conducted on glass showed that it was highly resistant to leaching. But, as the Pennsylvania State researchers pointed out in 1978, in most of the leachability tests, the waste form was treated with water at normal temperatures and pressures: "This test, however, is irrelevant after the waste has been buried—the pressures, temperatures and chemical environment will be different. . . ." When the researchers tested the glass under conditions that simulated actual burial, an entirely different result was obtained. "After one week the glass spheroid contained numerous fissures and several zones of coloration. After two weeks, the specimen had broken into several fragments and seemed to be totally altered."[26]

The conventional wisdom that salt is a viable disposal medium has also come under increased attack. Research has shown that salt contains trace amounts of water. Radioactive waste constantly gives off heat from radioactive decay, and as it warms the salt, water could migrate toward the heat source, surrounding the waste with a dense brine. "The mystique has built up that salt is dry and it's OK," David Stewart of the U.S.G.S. told the Washington Post. "Salt is not dry and it's not OK." According to Stewart, "Dense brines are corrosive environments, more corrosive than anything nuclear engineers have ever coped with."[27]

Another key concern in repository design is ensuring that the construction of the repository does not, in and of itself, make the repository site unacceptable. As a commissioner on the California Energy Resources and Development Commission told Congress in 1978, "the very act of emplacing the wastes in a formation compromises many of the desirable features needed for isolation," by "opening shafts and tunnels," by exposing rock to heat and radiation, and by replacing strong, stable rock with a weak, loose fill.[28]

An insufficient knowledge of geologic processes, technical problems, and the lack of a demonstrably suitable site still stand in the way of geologic disposal. Despite well-documented gaps in the knowledge base, some nuclear developers insist that safe waste-disposal methods already exist and that the real problem is the lack of the "willingness on the part of our nation's leaders to decide which solution is best for America." The nuclear power industry continues to claim in its advertisements and public-relations material that the mined repository concept *guarantees* that active wastes will be effectively removed from our environment *forever*."[29] The unthinking optimism of the nuclear mindset continues unchecked. The *experts* have learned little from the history of their failures.

PART SIX

Material Unaccounted For

In any case, whatever we do, let us heed some advice from George Orwell having to do with cant and hypocrisy and the effect of euphemistic and slovenly language on political thought.

Let us not say "sensitive" when we mean "dangerous"; or "strategic quantity" when we mean bomb quantity; let us not talk of "safeguards" when we mean occasional inspection of bomb material; let us not talk of becoming a "reliable supplier" again when we mean a willingness to sell dangerous material and equipment to any nation to whom it was offered in a more innocent time. Let us not talk of "potential nuclear explosives" when we mean "explosives." (Plutonium is no more a potential nuclear explosive than TNT is a potential chemical explosive.) Above all, let us not designate as "peaceful" . . . any activity that allows direct access to nuclear explosives. "Peaceful" is not peaceful unless we can protect it from use in bombs, unless it is safe.

Victor I. Gilinsky, Commissioner,
Nuclear Regulatory Commission
1980[1]

"Safeguard" is itself a misnomer, connoting more "safety" and "guarding" than is warranted. The system basically consists of information gathering and reporting—nothing more.

Senator John Glenn
1981[2]

Prior to Hiroshima, the fundamental physical principles underlying atomic bombs were understood by some scientists, but there remained doubt about whether such bombs could be constructed. The detonation of the atomic bomb at Hiroshima publicly revealed the most important atomic secret: nuclear bombs were not just a theoretical possibility. The problem for aspiring atomic-bomb builders changed from whether a bomb could be built to how to build one.

Today the principles of bomb design are widely understood. The necessary data on the behavior of nuclear materials are available in textbooks on nuclear engineering. A number of undergraduate students at American universities have already designed atomic bombs which experienced weapons specialists say would work.[3] Knowledge is no longer a serious barrier to the construction of a crude—but effective—atomic weapon.

Since 1945, everyone who has tried to detonate an atomic bomb has succeeded on the first attempt, as far as publicly available information shows.[4] The most important obstacle to building a primitive weapon is obtaining the requisite amount of *special nuclear material (SNM)*—either plutonium-239 or highly enriched uranium. Today, thirty-seven years after the Trinity test, a nation or group seeking membership in the *Nuclear Club* no longer needs an operation the size of the Manhattan Project.

The steps taken to protect SNM are known as *nuclear safeguards*. The goal of a nuclear safeguards system is twofold: first, to *deter* anyone from attempting to obtain nuclear materials for weapons manufacture by ensuring that there is a high probability that the attempt will fail; second, to make sure that if material were successfully *diverted*, the theft would be detected in a *timely manner*. Safeguards are needed at every nuclear facility which handles SNM, including fuel fabrication plants, plutonium reprocessing plants, weapons manufacturing plants, and reactors which use highly enriched uranium or plutonium for fuel.

Safeguard measures are divided into two categories: *physical security* and *material accountability and control (MAC)*. Physical security is concerned with controlling *personnel access* to weapons material; for example, protecting facilities handling SNM from armed assault or making it difficult to remove material from plant sites without authorization. It consists of enclosing plants and special nuclear materials within fences, walls, and vaults, setting up intrusion alarms, guards, and monitors, and checking all people and packages entering and leaving *nuclear compounds*.

Material accountability and control is concerned with tracking the *inventory* of SNM present in a facility. While this might sound like a straightforward procedure, it is actually a complicated and inexact task. Within a single plant, nuclear materials may undergo repeated chemical processing. Often the material will be dissolved, diluted, concentrated, chemically separated, and processed in solid, liquid, and gaseous forms. As it moves through the *process stream*, it will change form many times, mak-

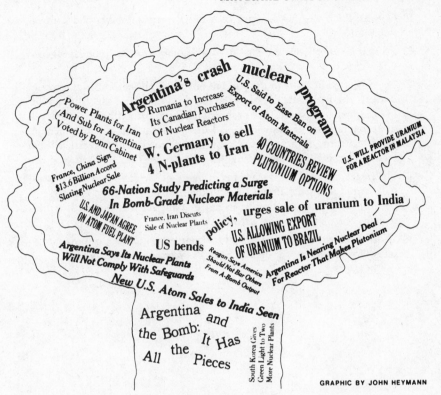

Argentina's crash nuclear program

Power Plants for Iran (And Sub for Argentina Voted by Bonn Cabinet

Rumania to Increase Its Canadian Purchases Of Nuclear Reactors

U.S. Said to Ease Ban on Export of Atom Materials

France, China Sign $13.6 Billion Accord Slating Nuclear Sale

W. Germany to sell 4 N-plants to Iran

40 COUNTRIES REVIEW PLUTONIUM OPTIONS

U.S. WILL PROVIDE URANIUM FOR A REACTOR IN MALAYSIA

66-Nation Study Predicting a Surge In Bomb-Grade Nuclear Materials

U.S. AND JAPAN AGREE ON ATOM FUEL PLANT

France, Iran Discuss Sale of Nuclear Plants

policy, urges sale of uranium to India

US bends

Reagan Says America Should Not Bar Others From A-Bomb Output

U.S. ALLOWING EXPORT OF URANIUM TO BRAZIL

Argentina Is Nearing Nuclear Deal For Reactor That Makes Plutonium

Argentina Says Its Nuclear Plants Will Not Comply With Safeguards

New U.S. Atom Sales to India Seen

Argentina and the Bomb: It Has All the Pieces

South Korea Gives Green Light to Two More Nuclear Plants

GRAPHIC BY JOHN HEYMANN

FIGURE 12 International trade in nuclear technology may be profitable for a few companies in the U.S. and Western Europe. But as the "peaceful atom" spreads, so does the capability for making bombs.

ing inventory difficult. No physical measurement system is perfect, and the repeated chemical processing injects additional uncertainties, all of which accumulate as the material runs through the process stream. Because the quantities of material involved are large, diversion of even a fraction of a percent of the total could provide enough material to construct a weapon.

A successful safeguards system must be able to *respond* to a wide *spectrum of threats*, ranging from a *covert diversion*, conducted by a well-organized group working in concert with a *knowledgeable insider*, to an *armed assault* on a nuclear shipment or facility. Each nuclear nation has its own *national safeguards* to protect its SNM. There is also a loose system of *international safeguards*, which are supposed to make it harder for a government to divert material from its own *peaceful facilities*, that is, to use *peaceful* nuclear technology as a *civilian cover* for a weapons program.

Safeguards specialists are confronted with a number of dilemmas that make their job very difficult, including:

☐ *Physical limits* on the accuracy of materials accountability and control systems.

☐ *Economic limits* on the amount of money that can be spent on safeguards.

☐ The projected increase in the amount of special nuclear material moving through the fuel cycle. As the *material flow* increases, the accuracy of MAC systems must increase proportionately to maintain the same level of accountability. If the widespread use of plutonium as a commercial fuel is sanctioned, the accuracy required to protect the material will be greater than is physically attainable.

☐ The instability of the governments of many of the nations which potentially could join the *Nuclear Club* in the next decade. Even if a present government is truly committed to not building nuclear weapons, a subsequent one might not share that view.

☐ The difficult problem of deciding how guarded is guarded, or how safe is safe? How does one decide a safeguards system is *good enough*?

CHAPTER FIFTEEN

No Evidence

The nuclear safeguards system of the 1960s and 1970s failed to keep track of special nuclear material accurately. Thousands of pounds of plutonium and highly enriched uranium turned up missing[5] and were designated *material unaccounted for*, or *MUF*. Despite strong circumstantial evidence that some of this material may have been stolen, nuclear developers have repeatedly claimed that there is *no evidence* of theft.

The case of the Nuclear Fuel Services Corporation's fuel fabrication plant in Erwin, Tennessee, illustrates the uncertainty surrounding MUF. In September, 1979, the NRC received a report from the Erwin plant saying that the facility had lost track of more than 9 kilograms (about 20 pounds) of highly enriched uranium, enough to construct at least one atomic bomb. The exact amount of uranium missing was kept secret.[6]

The plant's operating license stated that a loss of more than 9 kilograms of uranium required an immediate shutdown for a complete *inventory review*. Shortly after receiving the report, the NRC ordered the plant, which produces fuel for the Navy's nuclear submarines, temporarily shut down. This incident was only the most recent in a series of *inventory differences* at Erwin. Over the past decade, losses there exceeded 246 pounds.[7]

What happened to Erwin's missing material? There are a number of possible explanations for the inventory differences. SNM could have been trapped in the plant's *process stream*—for example, caught in pipes or tanks. It could have been lost in the plant's *waste stream*, disappearing into the air or into scrap and waste water. Missing SNM might never have existed: it could simply be the result of *prior miscalculation*. It could have been explained by *random measurement errors* or *systematic bias* in measurement devices. Or it could have been stolen. The purpose of the shutdown and reinventory at Erwin was to determine which of these explanations was correct.

At the Erwin plant, uranium is handled in liquid, solid, and gaseous form.[8] As the material moves through the process stream, undergoing chemical and physical changes, a great deal of uncertainty is injected into the measurement process. Moreover, as the output of each step in the process becomes the input of the next step, any error introduced moves along the process stream just as the material does. These errors accumulate and the uncertainty grows about how much material is where.

Determining whether an inventory difference is due to statistical *noise* in the measurement process or to theft or diversion is a difficult business. As William J. Dircks, head of the NRC's Office of Nuclear Material Safety and Security, told the journal *Science:* "We're not counting discreet items; we're really estimating the amount of uranium atoms that may be within the system at any one time. It's more like putting a net in a tank where you think you have a certain number of fish, and taking a statistical sample to see if you have the number of fish per meter that you think you should have. If the fish are all at the bottom feeding, or stuck somewhere [then you have a problem. Erwin is checking through all its miles of piping] to see where the fish are hiding."[9]

The NRC establishes *MUF thresholds*, or *alarm levels*, for each plant under its jurisdiction. (This includes commercial plants, but does not include federal installations, like Hanford, which are operated by the DOE and where reprocessing has been carried out since the Manhattan Project.) If the inventory difference exceeds the threshold—as it did in the case of Erwin— then an inventory review is conducted. But some critics charge that the reinventory process is nothing more than an elaborate charade. Thomas B. Cochran, a nuclear physicist with the Natural Resources Defense Council, a nonprofit environmental group, wrote in 1978,

> Under present NRC practices, the MUF game is played as follows: When the MUF exceeds [the alarm level] the facility is shut down. A big review is launched. Several possible *loss mechanisms* are "*discovered*" or "*postulated*," and then, casting all logic aside in one mystical leap, the NRC staff concludes the MUF is due to one or more of these loss mechanisms and there is "*no evidence* of diversion." Through this mechanism the MUF is "*explained*," i.e., reduced to less than [the alarm level] and the facility is allowed to reopen.[10]

In the case of the Erwin plant, however, the investigators were unable to develop a *plausible technical explanation* for the inventory difference; *satisfactory loss mechanisms* were not *identified*.[11] Dircks, who had been unable to determine where the fish were hiding, recommended that the NRC revoke the Erwin plant's license. But in January 1980, in a closed meeting, the commission voted to allow Erwin to reopen, easing the plant's *accountability requirements* by raising the *alarm threshold* to a level the plant could meet.[12] Since the plant could not meet the conditions of its license, the NRC simply relaxed the rules to make them "*more realistic.*" The DOE had sent the NRC commissioners a classified letter saying that the continued operation of the Erwin plant was essential to *national security*, apparently convincing the NRC to amend Erwin's license.[13] (In October 1980, after the plant had reopened with its new *relaxed license*, the NRC staff discovered new *loss mechanisms*, reducing the amount of MUF by more than half.)[14]

In February 1980, the Natural Resources Defense Council (NRDC) requested that a hearing be held on Erwin's license, and in July 1980, the NRC voted 3 to 2 to change its rules for holding adjudicatory hearings so that it would not have to comply with the NRDC's petition. The majority doubted whether the *"public interest"* would be served by holding a hearing since the case involved *"sensitive issues"* and "basic regulatory policy questions regarding the conduct of military functions."[15] One of the dissenting commissioners called the decision "dishonorable and disgraceful," charging that "the only thing being protected against here is the potential embarrassment to this agency [the NRC] or to the Department of Energy that might flow from effective probing of particular facts in this case."[16]

Throughout the Erwin controversy, the NRC repeatedly claimed that there was *"no evidence"* that the missing uranium had been stolen.[17] The repeated use of the phrase *no evidence*, a classic defense of the nuclear mindset, provoked NRC Commissioner Peter Bradford, who pointed out that the phrase "has historically been used in situations where normal English would have settled for a phrase more like 'no absolute proof.'"[18] Bradford cited the case of the Nuclear Materials and Equipment Corporation (NUMEC) in Apollo, Pennsylvania, to illustrate the importance of this linguistic distinction.

NUMEC produces highly enriched uranium fuel for the Navy's nuclear reactors and manufactures plutonium fuel-rod elements for the *Fast Flux Test Facility* (FFTF), an experimental breeder reactor at Hanford, Washington. Since the early 1960s, NUMEC has had a history of *inventory discrepancies* and security violations. As early as 1962, an AEC report noted that "numerous deficiencies were found in NUMEC's overall security program."[19] That report, however, was not made public until 1977.

While there have been literally hundreds of violations at the NUMEC plant, the most serious one was the discovery in 1965 that the facility had lost track of 381.6 pounds of highly enriched uranium. After inventory review, government officials decided that about half of the MUF had been lost through *"known mechanisms."* The remainder was still unaccounted for.[20]

Over the years, the nuclear agencies have publicly insisted that there is *no evidence* that the NUMEC MUF was stolen. But in secret documents and internal discussions, the government has shown less certainty about what actually happened. A secret memo to AEC Chairman Glenn Seaborg, written in 1967 and declassified in 1977, concluded that "it cannot be said unequivocally that theft or diversion has not taken place, but the *most probable explanation* is that NUMEC consistently underestimated its plant process losses. . . ."[21] Other secret documents, declassified at the same time, indicate that the AEC and the CIA suspected that China might have stolen the NUMEC uranium because atmospheric debris from the first Chinese bomb test resembled the missing NUMEC uranium. After the AEC concluded that China had developed its own source of weapons-grade uranium, this theory was discarded.[22] But suspicion that the uranium was stolen did not disappear.

In November 1977, the government declassified two documents written in 1976 which showed that the nation's intelligence agencies suspected that Israel diverted the NUMEC uranium. The documents, which were released under a Freedom of Information Act suit, indicated the FBI, the CIA, and the National Security Council all shared this view. No other country was named as the possible recipient of the material.[23] Nevertheless, before Congress and the media, ERDA and the NRC continued to claim that there was *"no evidence"* of diversion at NUMEC.[24]

As NRC Commissioner Peter Bradford pointed out in 1980, the *"no evidence"* position that the two agencies presented really described "a situation in which there was considerable circumstantial evidence, but no conclusive proof. When challenged, the defenders of this practice had tended to assert that evidence is synonymous with proof."[25]

By the summer of 1979, the NRC had stopped asserting that there was no evidence of a diversion at NUMEC. NRC Chairman Joseph Hendrie presented testimony before Congress in which he said that circumstantial evidence "points in one way, but not enough to go out and indict someone." His testimony was accompanied by a report which said that security at the NUMEC facility was so lax that "a knowledgeable insider could quite easily have obtained material in that plant."[26]

Commercial facilities like Erwin and NUMEC are not the only ones with MUF problems. The reprocessing plant at the Department of Energy's Savannah River Plant, a 300-square-mile nuclear complex which produces plutonium and tritium for nuclear weapons, had a *net shortage* of 145.5 kilograms of plutonium as of 1978.[27] In a report released in early 1980, the General Accounting Office criticized the DOE for concluding that there was *no evidence* that any of that MUF was diverted. According to the GAO report: "The Department of Energy assumes that none of this was diverted. It attributes the shortage to inaccurate production estimates, process measurements, shipper/receiver measurements, and accounting and normal operating losses. GAO believes that with existing material control and accountability technology, the Department has no valid basis for this assumption and is thus unable to provide definitive assurance that no plutonium has been diverted."[28]

CHAPTER SIXTEEN

The Game Theory
of Safeguards

As safeguards technology develops, it is approaching physical and economic limits that constrain further improvement. Yet if the use of special nuclear material continues to increase, safeguards will need to be more effective to provide the same level of protection. Safeguards theorists are developing a new strategy based on a *"game theory"* of safeguards. The problem is viewed as a contest between two *adversaries*: the *diverter*, who intentionally exploits weaknesses in the safeguards system, and the *defender*, who tries to thwart the diverter's efforts.[1] Just as nuclear war is studied by playing *war games* on computers in the Pentagon, safeguards strategy is being developed using game theory in the DOE's laboratories.

One safeguards specialist is William H. Chambers, deputy assistant director for safeguards research at the Los Alamos Scientific Laboratory, the nation's lead lab for material accountability and control research. (The nearby Sandia Laboratory, in Albuquerque, New Mexico, is the lead lab for physical security systems.) Interviewed in 1980, in an office on one of the *unclassified floors* of the National Security and Resources Study Center in Los Alamos, Chambers frankly discussed the difficulties inherent in protecting special nuclear material from theft. Throughout the interview, Chambers stressed that decisions about what is an *"acceptable risk"* of theft are essentially economic decisions: the risks of diversion must be balanced against the cost of additional protection. "People want *absolute assurance*," Chambers said. "We can't give them that. Just as we can't make automobiles 100% safe, we can't make nuclear materials accountability and control 100% perfect."[2]

Materials accountability and control technology has made substantial advances since the early years of nuclear safeguards. As one Los Alamos safeguards specialist put it, in the 1960s the industry "did not have the accounting system you see in a Coca Cola plant."[3] Despite the improvements, routine losses in today's reprocessing plants often run as high as 5% to 10% of total plant output.[4] Chambers and his colleagues at Los Alamos believe that with new and projected improvements in technology, this number could be reduced to about 1%.[5] But this improvement in accuracy would be dwarfed by the increase in the flow of plutonium that would result from the use of plutonium as a commercial reactor fuel. In the hypothetical world envisioned in 1977 by Alvin Weinberg, with 5,000 breeder reactors of 5,000 megawatts each operating, 30,000 tons of plutonium would be reprocessed annually, or about 100 tons per day.[6] One-tenth of 1% of this daily figure would, if diverted, be enough plutonium to make approximately ten bombs.

One of the main sources of uncertainty in monitoring the flow of plutonium through reprocessing plants is inaccurate analysis of the plutonium content of the spent fuel rods when they first come into the plant. Under present practices, the plutonium content of a spent fuel rod is estimated by the use of a mathematical formula with an average error of 10%. Since the amount of plutonium cannot be accurately determined at the outset, detecting a theft later in the process becomes problematic. Chambers said that "we are three to five years away from having an accurate fuel rod assay system."[7]

At this stage in its development, Chambers said, MAC technology is bumping up against fundamental *"physical limits"* that are constraining additional improvements in accuracy. He stressed that the measurement techniques that are used "are the best that are known," but that nevertheless, "error accumulates" and the best one can hope for is around a 1% margin of error.[8]

The increase in accuracy is not the only change in material accountability and control systems since the 1960s; the definition of *good accountability* has changed as well. In the old view, it was enough to conduct an audit of each plant's inventory every six months. More recently, the hazards of not knowing what was happening between audits were recognized, and the goal shifted to getting "more immediate information." At Los Alamos, scientists are developing instrumentation that may make it possible to assay the material stream in *real time*; that is, on a continuous basis, without shutting the plant down. This would allow accountability to be determined "in the course of a day's shift," Chambers said, allowing more *"timely detection of potential abnormalities."*[9]

But despite improvements in accuracy, instrumentation, and the science and art known as *diversion path analysis*, the MAC problem remains formidable. Most difficult is the problem of detecting *trickle diversions* that are within the *noise level* of the process. Such trickle (or *long-term*) diversions may never be detectable, for the simple reason that they would be totally obscured by the noise of the system.

Measurement uncertainty and other noise provide anyone planning a trickle diversion with considerable room in which to operate—even with the most stringent safeguards in place. As the GAO noted in its 1980 report on the proliferation risks of reprocessing technology, even a 1% measurement uncertainty "could result in as much as 150 kilograms of plutonium being unaccounted for per year" in a single commercial reprocessing plant.[10] That amount would be sufficient for construction of more than fifteen nuclear weapons. Protecting against trickle diversions may prove impossible. G. Robert Keepin, a Los Alamos safeguards specialist, said in 1980, "We can't meet the safeguards requirements on long-term diversion, and we may never be able to."[11]

Chambers said that using game theory, decision analysis, and computer modeling and simulation, it may be possible to detect some trickle diversions that are very near the *limits of detectability* by "looking for small *signals* within the noise."[12] If a number of errors that should be random seem to be tending in one direction, for example, it could indicate that a diversion was taking place. Though some advantage can be gained by the use of these mathematical and statistical techniques, it is unclear how much our ability to detect trickle diversions can be improved. Also, there is the possibility that a *knowledgeable diverter* who understands the principles underlying these techniques could develop effective *countermeasures*.

"We can't say that a safeguards system is 100% effective," Chambers said. "We'd like to say that, but we can't. It would be deceptive."[13]

Even with the best possible technology, MUF will never be eliminated. How much MUF is too much MUF? Chambers said that question "doesn't have a simple answer." Some people want accountability down to grams, he pointed out, but "in an industry which deals with metric tons, grams are a small quantity."[14] (One gram is equal to one-millionth of a metric ton, and no accounting system will ever approach the 99.999% accuracy required to attain accountability down to one part in a million.) The accounting system cannot be perfect, Chambers stressed, and what you consider to be good enough is a matter of what standards or criteria you set.[15]

One criterion that is frequently used relies on the concept of *trigger quantities*, *strategic quantities*, or *significant quantities*—the amount of material needed to construct a single nuclear weapon. The significant quantities criterion holds that a successful MAC system would ensure that no one could divert enough material to build a bomb without detection. The International Atomic Energy Agency, for example, defines a significant quantity as 8 kilograms of plutonium or 25 kilograms of highly enriched uranium. The IAEA also estimates that the *conversion time*, the period required to convert special nuclear material into a bomb, is in the range of one to three weeks. Under IAEA criteria, a safeguards system that is capable of detecting diversion of a significant quantity within the conversion time is considered *adequate*.[16]

This definition, however, is somewhat arbitrary. For one thing, a government or subnational group planning to assemble a weapon could obtain material from several different sources. Second, the amount of material required for the construction of a weapon is substantially less than what weapons designers call the *classical figure* for a critical mass—20 kilograms of highly enriched uranium. The critical mass is not a constant number but a variable quantity, dependent on the bomb design.[17] In effect, the size of a critical mass depends on the skill of the weapons designer. The actual limit, the smallest amount of material needed to construct a bomb, is classified. But it is substantially less than the classical figure. As former weapons designer Theodore Taylor said, it "takes much less—and how much less depends on how good you are at making bombs. . . . It isn't a matter of saying twenty and meaning eighteen."[18]

How proficient can a clandestine bomb maker be? A terrorist group with limited resources would do a less professional job of designing and constructing a bomb than a government seeking to quietly accumulate a nuclear arsenal. But then, one good engineer or physicist could make a tremendous difference in the quality of weapons produced.

Chambers said that systems analysis, game theory, and modeling techniques are also used in developing physical protection systems to restrict personnel access to special nuclear materials and to protect SNM from overt attack. While he noted that physical protection technology has "matured" more rapidly than any other MAC technology, he cautions that it still remains impossible to guarantee absolute security.[19]

"Any barrier can be penetrated . . . given an appropriate sized army," according to Chambers. "There's always some level of a threat that cannot be handled." Just as you cannot say that a material accountability and control system is perfect, he said, you cannot make a physical protection system that is perfect. What you consider "*good enough*," he said, "depends on your analysis of the *spectrum of threats*."[20]

Until recently, official analyses of the spectrum of threats showed a strong bias toward the low end. Representative Morris K. Udall, Democrat from Arizona, who presided over congressional investigations of nuclear materials security, said in 1976 that some facilities which handle special nuclear materials "are protected by as few as two guards armed with .38-caliber revolvers and shotguns."[21] This was the case despite the NRC's recognition, on paper, that an attack unit might consist of three or more well-armed, well-trained persons, equipped with explosives, machine guns, and antitank weapons. Security has since been beefed up, but the question of what is good enough remains.[22]

CHAPTER SEVENTEEN

A Nuclear-Armed Crowd

The word *peaceful* is one of the most abused words in the Nukespeak lexicon. The international safeguards system, which is supposed to prevent *peaceful nuclear materials* from being used for weapons, lacks both the technical and the political means to check nuclear weapons proliferation. *Peaceful nuclear technology*—and with it the capability to make bombs—is spreading rapidly, driven by the legacy of Atoms for Peace propaganda, the profiteering of nuclear exporters in the U.S. and Western Europe, and the desire of some nuclear importers to acquire the means for weapons production. India's atomic bomb was a byproduct of its *peaceful nuclear program*. Udai Singh, press secretary for the Embassy of India in Washington, D.C., insists that the bomb was a *"peaceful nuclear device*," saying that the development and explosion of the *device* was scientific *"research*." "It is a question of *intention*," Singh said; India is "not interested in producing nuclear weapons."[1]

Using their *peaceful* nuclear programs, a growing number of countries are developing the ability to make nuclear bombs (see figure 13). Since 1974, when India announced its membership in the *Nuclear Club*, the situation has grown more acute. By the end of the 1980s the world may see a sevenfold increase in the amount of plutonium produced in nuclear power plants: a total of more than 500 tons may be accumulated.[2] The nuclear technology involved in the planned shift to a plutonium-based fuel cycle is particularly *sensitive* to use for weapons purposes.

The International Atomic Energy Agency (IAEA), which promotes international nuclear power development, is charged with ensuring that governments do not use their nuclear programs to obtain material for weapons production. This assignment may prove impossible. The General Accounting Office concluded in 1980 that the "IAEA will have difficulty assuring the international community that a nation operating a commercial reprocessing facility is not diverting material for weapons purposes." One reason the GAO offered for its conclusion was that "the IAEA faces the same technical limitations . . . as are found in the U.S. safeguards system. . . ."[3]

The IAEA also faces political problems. Many nations considered likely to be interested in nuclear weapons have not ratified the international non-proliferation treaties through which safeguards are imposed. These include such *suspect states* as India, Pakistan, Israel, and South Africa.[4] And even if ratified, the treaties are ineffective. Iraq, for example, would not allow IAEA inspectors into its nuclear reactor during the Iranian-Iraqi War.[5] Israel considered the safeguards at the Iraqi reactor sufficiently weak to justify destroying the reactor in June, 1981. As Stanford economist Harry Rowen said, "Are we really going to have a bunch of people sitting around watching plutonium in Iraq, Pakistan, Taiwan, and South Africa? The safeguards people must be joking. Just think of the political realities."[6]

The desire to build nuclear weapons is contagious. Pakistan, for example, called India's nuclear test "a threat" and a "fateful development."[7] (The Pakistanis were not impressed by India's suggestion that they were overreacting.)[8] In 1979, the U.S. concluded that Pakistan had begun a nuclear weapons program.[9] The Indian threat was probably an important motivation for Pakistan to *go nuclear*.

As a result of the proliferation of both the means and the will to develop nuclear weapons, the *Nuclear Club* is becoming what some arms control specialists have termed a *nuclear-armed crowd*.[10] Nuclear weapons may *proliferate horizontally* to dozens of nations in the next two decades. Meanwhile, *vertical proliferation*—the expansion of the nuclear arsenals of the superpowers—remains unchecked.

A nuclear-armed crowd would pose new and dangerous threats to peace. One study of the military potential of *peaceful* nuclear programs concluded: "The rather tired arguments about *perfect safety* through vastly increased *mutual vulnerability* have not worn very well in the case of a hypothetical world of two countries and of *rationally calculating men*. *Universal vulnerability* in a world of many nuclear powers including perhaps Khadafi [the ruler of Libya] is even less attractive."[11]

While the threat of proliferation is widely recognized, there is a growing sense that nothing can be done about it. This fatalistic attitude could become a self-fulfilling prophecy: some nuclear exporters are arguing that the few proliferation controls now in effect should be lifted since the spread of nuclear weapons is inevitable.[12] Yet as advocates of stronger controls have pointed out: "A fatalism which holds that nothing can be done today may be an unconscious cover for a desire to do nothing, to continue as before. This would be intelligible in light of the large vested interests, as much psychological as financial, in the movement towards the use of plutonium fuel."[13]

NRC Commissioner Victor Gilinsky warned about this fatalism in 1980:

We are talking about bombs, nuclear bombs with which India and Pakistan can incinerate South Asia; bombs with which Iraq or Libya can subju-

Potential for Nuclear Weapons Proliferation

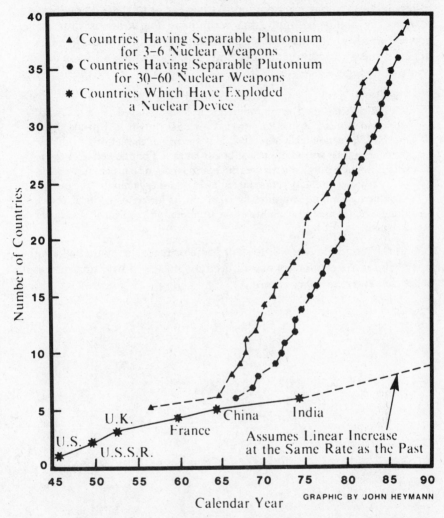

FIGURE 13 The potential for nuclear weapons proliferation. As nuclear power technology spreads around the globe, so does the ability to make nuclear bombs. In the past, nuclear weapons have proliferated at a linear rate. But because of allegedly peaceful nuclear programs, the number of countries with enough plutonium to manufacture nuclear weapons is now rising very rapidly.

Source: Adapted from Albert Wohlstetter et al., *Swords From Plowshares* (Chicago: University of Chicago Press, 1979), p. 15.

gate Saudi Arabia or set the whole Middle East aflame and shut off *everybody's* oil. The threat of nuclear weapons in the Middle East, or Pakistan, or South Africa does not mean we should back off our nonproliferation policies. It means just the opposite. (original emphasis)[14]

The inadequacy of safeguards technology and institutions like the IAEA fosters fatalism about proliferation. As nuclear power technology spreads, so does the ability to apply that technology for weapons. But nuclear power—and with it, unavoidable proliferation—is not the only energy option. As energy strategists Amory and Hunter Lovins and Leonard Ross wrote in *Foreign Affairs* in 1980,

> The nuclear proliferation problem, as posed, is insoluble.
>
> All policies to control proliferation have assumed that the rapid worldwide spread of nuclear power is essential to reduce dependence on foreign oil, economically desirable, and inevitable; that efforts to inhibit the concomitant spread of nuclear bombs must not be allowed to interfere with this vital reality; and that the international political order must remain inherently discriminatory, dominated by bipolar hegemony. . . . These unexamined *assumptions* artificially constrain the arena of political choice and maximize the intractability of the proliferation problem. . . . (original emphasis)[15]

In the next part, we will look at some of these "unexamined assumptions" about the economics of and need for nuclear power and dissect the belief that *there isn't any choice*.

PART SEVEN

There Isn't Any Choice

Systematic confusion of expectation with fact, of hope with reality, has been the most characteristic feature of the entire 30-year effort to develop nuclear power.

Irvin C. Bupp
Jean-Claude Derian[1]

There are only seven percent [actually, five and a half percent] of the people of the world living in the United States, and we use thirty percent [actually, thirty-three percent] of all the energy. That isn't bad; that is good. That means we are the richest, strongest people in the world, and that we have the highest standard of living in the world. That is why we need so much energy, and may it always be that way.

President Richard M. Nixon, 1973[2]

The euphoric vision of a nuclear-powered *Garden of Eden* became the blueprint for postwar U.S. energy policy. In the late 1940s and early 1950s, energy use grew exponentially, increasing at the rate of 3% per year or more. Electricity use grew even faster, doubling every ten years. To people who examined the data, it was obvious that America's supplies of cheap oil and gas would not be able to sustain this growth rate forever. Without a change in energy policy, the U.S. would become dangerously dependent on foreign energy supplies by the 1970s.

There were two possible responses to this situation. The first was to rely on achieving higher energy efficiency, thus curbing exponential energy growth. The second approach was to introduce nuclear power rapidly to maintain exponential energy growth.

The U.S. government chose the second approach, embracing untested nuclear technology as the solution to the energy problem. Nuclear prophets assumed that continued exponential growth of energy use was possible, desirable and necessary. They assumed that nuclear power could supply ever rising quantities of energy. They assumed that the power of the atom could be harnessed easily, that safety questions would be quickly resolved, that hundreds of nuclear plants could be constructed rapidly, and that nuclear electricity would be cheap. They assumed that no other energy source would be capable of meeting people's energy needs. Even before there were any working nuclear power plants, they presented these assumptions as *facts*. They drew up lavish plans, based on speculation and hope, for a world powered by thousands of breeder reactors. Heady from the scientific and engineering successes of the Manhattan Project and backed by the political might of the Atomic Energy Commission, nuclear developers rarely doubted that nuclear power was the technological panacea they sought.

But while the vision of a nuclear future was politically powerful, nuclear power technology proved to be weak. Electricity from nuclear reactors is not *too cheap to meter*; even with huge federal subsidies it remains expensive. The growth of the nuclear industry has consistently lagged behind official projections. In the 1970s, the rising costs of nuclear construction, combined with reductions in the electricity-use growth rate, led to the cancellation of many proposed reactors. Sales of new reactors have slipped to such a low level that the nuclear industry is in danger of complete financial collapse.

Nevertheless, many nuclear developers have not lost faith in nuclear power. Since the 1973 oil embargo, however, the vision of the nuclear future has changed. Nuclear power is no longer hailed as the harbinger of a new age of abundant and cheap energy; instead it is presented as the only hope. Nuclear developers have repeatedly argued that we have *no choice*, that unless we wholeheartedly adopt nuclear expansion we will wind up *freezing in the dark*.

The belief that there is no choice but to rely on nuclear power rests on the assumption that energy growth is necessary to sustain economic growth. As Thomas A. Vanderslice, senior vice-president of General Electric Company, wrote in *Public Utilities Fortnightly* in 1979, electric utilities "are going to need a lot more capacity because the people of this country want at least moderate economic growth and that is going to mean more energy."[3] Though the view that Vanderslice expressed is widespread in the energy industry, it is not correct. Energy-efficiency techniques can reduce energy consumption without halting economic growth. An energy strategy based on increased efficiency in energy use and a transition to solar energy sources can support a productive and healthy economy. The barriers to implementing such a strategy are not technological; they are a result of the nuclear mindset. By draining huge amounts of capital that could otherwise be invested in a transition to an economy based on efficient use of renewable solar sources, nuclear power is prolonging the energy crisis.

CHAPTER EIGHTEEN

Soft Numbers and Hard Times

Nuclear developers planned a huge nuclear power program. In 1974, the Atomic Energy Commission projected that the U.S. would have between 850 and 1,400 nuclear reactors in operation at the turn of the century.[4] Such proposals were based on faith in nuclear power technology rather than on data about its costs and performance. In the absence of hard information, nuclear developers made euphoric predictions about the price and availability of nuclear energy. National and international energy policy was based on "soft" numbers.

A large nuclear power program was seen as the way to sustain increases in energy consumption. In the late 1940s, oil and gas reserves were being depleted rapidly, and energy use was growing. If the billions of people in the non-industrial world were to make a transition to energy-intensive economies, fossil-fuel use would rise even faster. Analyses of oil reserves suggested that consumption was already increasing so fast in the U.S. that the country might burn up most of its cheap oil before the end of the century.

In 1949, M. King Hubbert, a Shell oil geologist, published a paper warning that fossil-fuel supplies were likely to be short-lived. Hubbert's paper showed that the curve describing the growth of oil consumption was headed nearly straight up. Since oil reserves were finite, the faster the rate of increase in their use, the sooner they would be exhausted. Eventually, oil production would peak, and the economy would come tumbling down the backside of the curve.[5]

Hubbert was not saying that every last drop of oil would be pumped or that every last seam of coal would be mined. There would always be more oil and coal out there somewhere. But in a market economy, it was inevitable that the easy-to-find, easy-to-extract resources would be used up first. It was equally inevitable that energy prices would rise rapidly as energy producers turned to ever more exotic sources, such as offshore oil, oil shales, tar sands, and synthetic liquids and gas from coal. Hubbert emphasized that a civilization dependent on ever increasing supplies of cheap energy would be in trouble when the cheap sources began to run out.

Nuclear developers believed that nuclear energy had arrived just in the nick of time to allow the exponential growth of energy use to continue into the twenty-first century. They had little doubt that as the fossil fuels were exhausted, nuclear power would be there to take up the slack, supplying ever increasing amounts of electricity at ever decreasing prices.

In 1949, the AEC commissioned Palmer Putnam, a consulting engineer who had done some pioneering work on the wind generation of electricity, "to make a study of the *maximum plausible* world demands for energy over the next 50 to 100 years."[6] Putnam was to pay special attention to the "maximum plausible demand for nuclear fuels" to aid the AEC in planning the development of nuclear power.[7] In carrying out this charge, Putnam took the position of a "hypothetical Trustee" of the world's energy future in his 1953 book, *Energy in the Future*.[8]

Putnam's historical analysis showed that world energy use was increasing dramatically. Between 1850 and 1950, the world's people had consumed "at least half as much energy as in the preceding eighteen and one-half centuries."[9] For most of the past two thousand years, world energy use had been low, averaging no more than about 4.9 quads per century.[10] (A quad is a unit of energy equivalent to one quadrillion British thermal units. In 1979, the United States alone used 79 quads of energy.) By 1850, the world rate of energy consumption had reached an average of 1,000 quads per century. By 1950, the rate had soared to 10,000 quads per century and was still climbing at a rate of 3% per year.[11]

Putnam argued that the standard of living could not continue to rise unless larger and larger amounts of low-cost energy were available. Sustaining exponential energy growth became the centerpiece of his analysis. To keep this growth going, energy sources that were both abundant and cheap needed to be found. Putnam recognized that if the price of energy began to increase sharply, energy growth would fall: "If, as we suppose, low-cost energy is essential to a lively economy, then, we argue, energy costs higher than 2 times 1950 costs would not permit us to continue to *live high*. They would act as a brake on the economy. The hypothetical prudent Trustee would wish to do all possible to avoid such cost increases."[12] Fossil fuels were woefully inadequate for the task. Nor did Putnam think that increases in the efficiency of energy use could keep up with the rise in the demand for energy.[13] He also examined solar energy as a possible "income" source to replace our reliance on the "capital" fossil-fuel sources. For Putnam, solar energy included fuel wood, farm wastes, water power, wind power, solar heat collectors, fuels from cultivation of special algae, and temperature differences in tropical ocean waters. His predicted energy demand in the next century was so large, however, that solar energy fell short,[14] even though his estimates of solar's potential contribution were very large in comparison to the amount of energy we now derive from solar sources.

In the absence of any new low-cost energy sources, Putnam believed that continued exponential growth in world energy use would have to be abandoned. To avoid this, he threw nuclear power into the breach. Despite the total lack of any experience in operating nuclear power plants among the people of the world, Putnam assumed that nuclear power would meet his criterion of supplying power at no more than twice the 1950 cost of coal. He also assumed that solar power would cost forty times as much as coal, making nuclear power by far the preferable way to go.[15]

Nuclear power makes its major contribution in the form of heat to generate electricity. Without a major restructuring of the economy, Putnam estimated that nuclear-generated electricity could meet only 15% of the "total maximum plausible United States energy system in A.D. 2000."[16] Putnam thought that nuclear power was sufficiently attractive as a long-term energy source that it might justify expanding the electrification of the economy to the point where nuclear power could supply at least 60%, rather than 15%, of the total energy. This change would not be cheap, but Putnam argued that it would be cheaper than trying to derive the equivalent amount of energy from solar power, given his high-cost assumption for solar as compared to coal.[17]

Putnam concluded his study with a call for action to accelerate the development of nuclear power. Assuming that electrification could be greatly expanded, Putnam said that "nuclear fuels could support the bulk of the maximum plausible energy systems of the United States and the Free World for several centuries." Putnam then had the hypothetical Trustee urge that "the nation's talents, public and private, be released" for developing *"nuclear furnaces"*; that we explore the possibility of harnessing energy from fusion reactions; and that we begin searching for new sources of the exotic metals needed for a large nuclear industry. The Trustee had one final note of caution, the last sentence of the book: "Finally, as our ultimate anchor to windward, he would urge the exploration of all ways to obtain income energy from sunlight in more useful forms and at lower costs than now appear possible."[18]

The AEC completely ignored this last bit of advice about developing solar energy. Instead, it made a sharp increase in electrification the heart of its response to the fossil-fuel problem. Exponential energy growth could continue, with nuclear-generated electricity substituting for functions then performed by fossil fuels.

The AEC began investigating several different reactor designs, but two styles quickly became dominant: *light-water reactors* and *breeders*. The military pressure to build a compact reactor for submarines led to the rapid development of the pressurized-water reactor. Once this technology was demonstrated, the AEC and the Joint Committee on Atomic Energy seized on it for the commercialization of civilian electric power.

Light-water reactors were regarded as a transitional technology, filling the energy gap until the breeder could be developed. As Alvin Weinberg noted in 1978, most nuclear developers "always believed that the breeder is the *essence of nuclear energy. . . .*"[19] Nuclear developers were drawn to the breeder reactor because without it, there could be no long-term nuclear fission future. The light-water reactors in use today utilize only 0.6% of the energy content of uranium, since they burn only the rare isotope, U-235. According to a 1978 article by Peter Fortescue of the General Atomic Company, light-water reactors could supply a maximum of 1,800 quads of energy using U.S. ore reserves. Since the U.S. energy consumption in 1979 was 79 quads, light-water reactors were obviously not a long-range energy source. Fortescue also estimated that burning all of the U.S. estimated coal reserves would produce 4,500 quads, and burning all of the world's estimated coal reserves would produce 18,000 quads.

Breeder reactors could make much better use of the energy content of uranium by converting U-238 into fissile plutonium, which could then be burned. Fortescue estimated that breeder reactors could provide 360,000 quads of energy, enough energy to "assure our *most exorbitant needs* for all the future upon which we are entitled to speculate."[20]

Thoughtful breeder proponents, such as Alvin Weinberg, recognize that a nuclear-powered world would have unique problems. In 1977, as a kind of thought experiment, Weinberg imagined a world powered by 5,000 breeder reactors, each supplying 5,000 megawatts of electricity, five times the amount produced by a typical light-water reactor of today. He concluded that such a system would be "*plausible*" because the "*readily calculable constraints* do not appear to be limiting." Weinberg's thought experiment illustrated the way the scope of the problems posed by nuclear technology grows as the scale of the industry does. In a 5,000-breeder-reactor economy, there would be 125,000 tons of plutonium, and 30,000 tons would be reprocessed every year, or about 100 tons a day, an amount Weinberg described as "staggering." This huge quantity of plutonium would pose an intractable *trickle diversion* problem (see chapter 16).

Accident risks would also be troublesome. There are no Rasmussen-style estimates of the probability of core meltdowns in breeder reactors. But using Rasmussen's estimates for uncontained meltdowns in light-water reactors, Weinberg calculated that an uncontained meltdown would occur on the average of once every four years. Weinberg conceded that such an accident rate would not be "*tolerable* . . . in the present climate." But things could change. Surely, he argued, "it is *fair to assume*" that the accident probability would be reduced. More important, "the public will eventually accept radiation as *a part of life's hazards* rather than being viewed as something mysterious and special."

Weinberg's thought experiment reinforced his view that "the price nuclear energy demands, if it indeed becomes the dominant energy system, may be an attention to detail, and a dedication of the *nuclear cadre* that goes much beyond what most other technologies demand. I realize that many in our nuclear community would deny these assertions: but I would insist that we are unaccustomed, perhaps unwilling, to project our technology as far as I have—unwilling, in a sense, to face up to the *consequences of complete success.*"[21]

The U.S. breeder-reactor program has been a complete failure. Breeder proponents assumed that breeders would work in practice as well as in theory and that they would supply cheap power. Nuclear developers from the AEC, ERDA and the DOE have committed billions of federal dollars to the search for a breeder future, but a successful commercial breeder has never existed in the U.S. The only commercial breeder ever operated in this country was the Enrico Fermi I, which suffered a partial core meltdown and never operated profitably (see chapter 9).

The unwarranted optimism of breeder proponents is characteristic of the nuclear mindset. Since the beginning of the nuclear power program, nuclear developers have frequently confused hope with reality.[22] In *Light Water: How the Nuclear Dream Dissolved*, an analysis of nuclear economics published in 1978, Harvard Business School professor Irvin C. Bupp and French nuclear analyst Jean-Claude Derian described the way government and industry both failed to question the euphoric estimates of ever falling nuclear costs. There was "a circular flow of mutually reinforcing assertions that apparently intoxicated both parties and inhibited normal commercial skepticism about advertisements which purported to be analyses. . . . what was missing during this entire period of intoxication was independent analysis of actual cost experience."[23]

These intoxicating assertions got a major boost in 1963, when the reactor manufacturers sold *turnkey reactors* to promote the development of nuclear power. Manufacturers offered to build these plants at prearranged fixed prices, with the manufacturer absorbing any cost overruns that might occur. The prices quoted for these plants, which later turned out to be well below the actual costs, appeared to be competitive with coal-fired plants. With the manufacturers carrying the risk, utilities decided to *go nuclear*. U.S. reactor manufacturers then used the sale of turnkey plants to persuade European governments to adopt American reactor technology.[24]

By the mid-1960s, utilities were ordering reactors at a rapid pace. In 1966 Alvin Weinberg and Gale Young from Oak Ridge gave a lecture on the exciting prospect of *"The Nuclear Energy Revolution."* Utilities had orders for twenty-nine nuclear power plants, and Weinberg and Young predicted a glorious future: "Nuclear reactors now appear to be the *cheapest* of all sources of energy. We believe, and this belief is shared by many others working in nuclear energy, that we are only at the beginning, and that nuclear energy will become cheap enough to influence drastically the many industrial processes that use energy. If nuclear energy does not, as H. G. Wells put it in 1914, create 'A *World Set Free*,' it will nevertheless affect much of the economy of the coming generation."[25]

The AEC was so certain that nuclear power would be cheap that in the summer of 1968 it sponsored a conference on *"Abundant Nuclear Energy"*

1968 AEC Forecast of Declining Electricity Prices

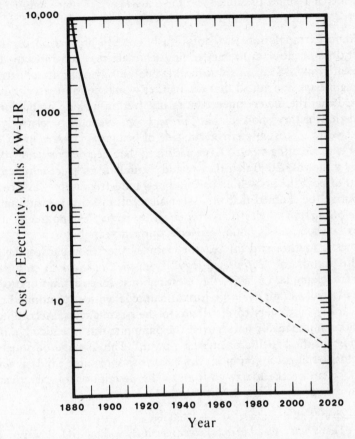

GRAPHIC BY JOHN HEYMANN

FIGURE 14 AEC forecast of declining electricity costs. At a 1968 conference on "Abundant Nuclear Energy," AEC researchers predicted that nuclear power would allow electricity prices to continue their historical decline into the twenty-first century. Shortly after this conference, electricity prices—including the price of nuclear-generated electricity—turned upward, and are still climbing in 1982.

Source: James A. Lane, "Rationale for Low-cost Nuclear Heat and Electricity," in *Abundant Nuclear Energy,* proceedings of a symposium, 26–28 August 1968 (U.S. Atomic Energy Commission, Division of Technical Information), p. 8.

to make plans for altering the nation's chemical and metal industries to exploit the electricity bonanza. At the conference, James A. Lane predicted a dramatic decline in the cost of electricity (see figure 14). Lane was an advocate of using the trends of the past to predict the costs of the future:

> How then does one take into consideration the combined effects of foreseeable and unforeseeable advances in technology on future power costs? The

answer, in my opinion, is to look back at historical trends that include both factors and simply to assume that these trends will continue into the future.[26]

Trend extrapolation is a notoriously superficial method of analysis, since it does not look at the underlying structure that generates the trends. In the early fall of 1929, a stock-market analyst relying on trend extrapolation might have concluded that the market would continue to climb. Lane admitted that there were uncertainties involved in making predictions, but argued that history showed that *"projections*, even for a relatively short span of years . . . usually err on the side of being *too conservative*."[27]

By extrapolating trends, Lane predicted that electricity prices would be so low by the year 2000 that they would "stimulate an *'electricity binge'* on the part of both the household user and new industrial users." In a concluding flourish, he claimed that the "over-all result may lead to fulfillment of the *age-old dream* of electricity *too cheap to meter*,"[28] apparently a reference to Lewis Strauss's famous speech fourteen years before.

Electric utilities and the reactor manufacturers bombarded the public with the promise of "*Infinite Energy*" from the "*Nuclear Energy Revolution*." For example, in 1967, the Westinghouse Electric Corporation produced a visually stunning, large-format pamphlet entitled "*Infinite Energy*" for its utility customers to distribute to their rate-payers. According to a small note on the inside back cover, the pamphlet was intended "*to inform the public* of the benefits of nuclear power." The 48-page document featured glossy, full-color graphics, with a cover shot of some children whirling around on an ornate merry-go-round. The description of the cover proclaimed that nuclear fission

> will give us all the power we need and more.
> That's what it's all about. Power seemingly without end. Power to do everything man is destined to do. We have found what might be called *perpetual youth* . . . and like the children on the cover . . . will have the hope and exuberance of *boundless energy* [ellipses in original].

Westinghouse was pulling no punches with its promise of "infinite energy." Inside, one section explained "What Infinite Energy Will Mean to You and Your Family." This 1967 vision of the all-nuclear-electric future was as uninhibited as R. M. Langer's back in 1940 (see chapter 3). Our homes would have movable roofs that could "roll back automatically to flood them with fresh air, sunshine, or moonlight" whenever we tired of the "completely controlled" indoor climate. Farms would be brought into the cities in indoor, "multi-level structures." There would be no farmers or farm workers as we now know them: "Computer-operated machines will plant, fertilize, and harvest crops." We could grow whatever food we wanted anywhere, such as bananas and coffee in Alaska.

Westinghouse promised that we "will soon get practically all the best things in good living by putting a plug into a wall, pushing a button, or flicking a switch." Whatever the problem, "all the best answers seem to come from electricity." Worried by air pollution? Try a nuclear power plant. Hate traffic jams? What about "high-speed electric trains; fantastic new electric urban skybus systems; pollution free electric cars"?

According to Westinghouse, it was "already evident" that nuclear power plants "will *create an age* of almost limitless electric power... *limitless power* will solve some of today's most pressing problems...."[29] These wild projections were based on extremely limited experience with the actual construction and operation of reactors. The eagerness of reactor manufacturers to forecast "limitless power," and of utilities to invest billions of dollars for such an untested technology, is an example of the hallucinogenic powers of the nuclear mindset. By the end of the 1960s, utilities were ordering nuclear plants six times as large as the largest plant then in operation.[30] James Lane at Oak Ridge in 1968 expected the "maximum unit size" for reactors to reach 1,500 megawatts by 1975, 2,000 megawatts by 1980, and 10,000 megawatts by 2010.[31] (By 1981, no reactor in the United States had reached the 1,500-megawatt size, and most plants on order were closer to 1,000–1,200 megawatts in size.) Nuclear developers believed that this rapid scaling-up in size would not only not cause problems, but would actually lead to the lowering of costs through economies of scale.

In less than ten years, these visions of the *"Nuclear Energy Revolution"* and *"Infinite Energy"* have been crushed under the ever mounting cost of nuclear power plants. The average light-water nuclear plant ordered in the mid- and late 1960s cost more than twice as much to complete as was originally estimated, after adjusting for inflation.[32] On the average, plants entering service in 1975 were about three times more expensive than commercial plants completed five years earlier, adjusting for inflation.[33] Nuclear-plant construction costs continue to escalate today at rates higher than the general inflation rate.[34]

The cost of constructing power plants is only one of the factors which determine the cost of the power they produce. Fuel costs must also be taken into account. Nuclear developers maintain that the price of nuclear electricity is less than that of coal- and oil-generated electricity because nuclear fuel costs are lower. Yet nuclear power, like fossil-fuel energy sources, is the beneficiary of large federal subsidies that make its true price difficult to compute. Throughout this century, energy prices have been distorted and manipulated through price controls, tax subsidies, price fixing, monopolies and cartels. In a free-market economy, prices are supposed to give consumers accurate information. If buyers are to choose correctly, then prices must accurately reflect the costs of commodities. Consumers rarely have a chance to make accurate comparisons between energy prices.

The nuclear industry has received billions of dollars in subsidies, including government support of research and development, uranium-enrichment, foreign reactor sales, and the deferred cost of nuclear waste disposal. Estimates of the total amount of these subsidies are very soft numbers. The development of nuclear weapons and the development of nuclear power have been so intimately intertwined that in many cases it is difficult to decide whether a given expenditure should be credited to the weapons program or to commercial reactor development. Since the amounts of money involved are large, differences in accounting practices will produce big differences in the estimates. Some nuclear developers have even claimed that the industry has received *no subsidies*.[35] In 1979, the General Accounting Office estimated that federal subsidies of the commercial nuclear industry have cost taxpayers $12.1 billion since 1950.[36] A 1981 report by the DOE's Energy Information Administration estimated that the domestic nuclear power industry has received $12.8 billion (1979 dollars) in federal support.[37] An earlier draft of the study, which was leaked to the press in 1980, estimated that commercial nuclear power had received $37 billion in federal subsidies since 1950. Without these subsidies, the draft report estimated, nuclear electricity would probably cost between one-and-one-half to two times its current cost. (The draft report was titled "Federal Subsidies for Nuclear Power: Reactor Design and the Fuel Cycle." In the final report, the word *subsidies* was deleted from the title and replaced by the word *support*.)[38]

Tax subsidies have also provided additional billions of dollars in hidden federal money to the nuclear industry, subsidies that were not included in the above estimates. An article in the *Harvard Environmental Law Review* in 1980 detailed the tax loopholes that benefit the electric utilities, noting that the benefits of tax subsidies are "especially significant in the case of nuclear power." The article pointed out that between 1954 and 1976, "total electric utility operating revenues increased 664%; net income increased nearly 400%. During this same period federal income tax payments by these utilities *decreased* 31%" (original emphasis).[39]

While the data on the true cost of nuclear electricity are very soft, the nuclear industry has produced misleading cost data to back up its hard sell of nuclear power. For example, in May 1979, the Atomic Industrial Forum (AIF), an industry trade association, released a *"survey"* of the costs of generating electricity from nuclear power and coal, based on plant operations during 1978. The AIF understated the cost of nuclear power and overstated the cost of coal. The survey used the kinds of deceptive techniques that have given statistics a bad name, providing a perfect example of devious mathematical manipulations like those documented in Darrell Huff and Irving Geis's 1954 book *How to Lie With Statistics*.[40]

The Atomic Industrial Forum distributed the results of its survey in an *AIF INFO* news release on May 14, 1979. The release announced that "forty-three of the 48 nuclear utilities responded to the survey. They reported that a nuclear kilowatt-hour of electricity cost, on average, 1.5 cents to produce in 1978, about the same as in 1977 and 1976. A coal-generated kilowatt-hour cost 2.3 cents in 1978, up from 2.0 cents in 1977 and 1.8 cents in 1976."[41]

A 1980 report by economist Charles Komanoff revealed that the AIF survey contained built-in biases. The AIF distorted the data by carefully selecting the plants used in the survey. The 21 reactors not included cost an average of 60% more to construct and produced 19% less electricity than the 39 reactors in the survey. Only 2 of the 14 most expensive reactors were chosen, and only 1 of the 7 reactors which operated at less than 50% capacity in 1978 was included. Komanoff calculated that the odds against randomly arriving at this peculiar selection of plants were 75,000 to 1.

The survey also excluded coal-fired plants operated by utilities with omitted nuclear plants. This twist eliminated American Electric Power and the Tennessee Valley Authority, which together produced more electricity from coal than all the coal-fired plants in the AIF survey. American Electric Power's coal cost was 1.75 cents per kilowatt-hour, and TVA's was 1.67 cents per kilowatt-hour, well below the AIF's reported figure of 2.3 cents.

After adding in the reactors and coal plants the AIF left out, Komanoff calculated that nuclear had cost 1.9 cents per kilowatt-hour, while coal had cost 2.1 cents per kilowatt-hour. With the missing data added, nuclear had cost 10% less than coal rather than the 34% less the AIF reported. Komanoff suggested two additional corrections: increasing the estimated cost of nuclear waste disposal and dismantling old reactors, and making an adjustment to reflect the cost of coal more accurately. With these changes, Komanoff calculated a nuclear cost of 2.0 cents per kilowatt-hour and a coal cost of 1.9 cents per kilowatt-hour. With these corrections nuclear had cost 7% more than coal.

Komanoff stressed that figures from simple surveys like the AIF survey should not be used "as guideposts for future costs," even with corrections. New plants cost more to build than old plants did, for example, and any survey that fails to distinguish between old plants and new ones will not predict future costs accurately. Komanoff said he did not offer his corrected figures "as a definitive basis for cost comparison between nuclear and coal," but because "they illustrate AIF's willingness to manipulate data in an attempt to portray nuclear power in an unjustifiably favorable light. . . ."[42]

Komanoff's study demonstrated that the AIF survey had been grossly misleading. When reporters confronted the AIF with evidence of the distortions, an AIF spokesperson said: "We stand by our survey, which is exactly what it is—*a survey*. It does not pretend to be a *definitive study* of the economics of nuclear power."[43]

By the time that Komanoff had discredited the AIF's dishonest survey, the survey's results had been widely published. The 1979 version of the

Edison Electric Institute's pamphlet "Nuclear Power—*Answers to Your Questions*" reported that

> industry statistics also *demonstrate* that nuclear power plants are more economical throughout the nation. The Atomic Industrial Forum reported that the *average total costs* of a kwh of electricity produced by various fuel sources during 1978 were:
> 1.5 cents/kwh for nuclear
> 2.2 cents/kwh for coal
> 3.5 cents/kwh for oil[44]

A June 13, 1979 report of the General Accounting Office on "Nuclear Power Costs and Subsidies" said that while 1978 figures on nuclear power costs were not available from the Department of Energy, "the Atomic Industrial Forum reported that 1978 generating costs were 1.5 cents per kilowatt-hour, about the same as they were in 1977 and 1976."[45] The AIF's doctored statistics turned up again in the February 15, 1980 issue of *Science* in an article by David Bodansky, chairman of the physics department at the University of Washington, who cited them as "*nationwide average ... costs.*"[46]

Whatever the cost of nuclear power may be, it is much higher than nuclear developers predicted. The cost of nuclear power is not falling; it is rising. Moreover, the electric power industry overestimated the demand for electricity. Until the mid-1970s, utilities based their expansion programs on the assumption that electricity use would grow at the rate of about 7% per year. This forecast was a straight-line continuation of the historical trend of electric demand for most of the century. Nuclear power would help make this growth possible by providing cheaper and cheaper electricity.

But nuclear power was not cheap, and electricity prices were not destined to continue falling. In the early 1970s, responding to the peaking of U.S. oil production and the OPEC oil embargo of 1973, utilities began to raise their rates, and consumers began to moderate their demands for more electricity. Demand growth fell to 0% in 1974 and has not returned to the 7% level since.[47]

The soaring costs of nuclear construction, coupled with the decline in energy growth rates, led many utilities to cancel or delay the construction of proposed power plants (see figure 15). The General Accounting Office reported that the country's electric utilities had canceled 184 planned power plants during the years 1974 to 1978—including 80 nuclear and 84 coal-fired plants. In addition, almost all the remaining power plants suffered construction delays, and the length of the delays increased. The average delay for nuclear plants during the years 1974 to 1978 was 33 months, more than three times longer than the average delay for coal plants. By 1978, the length of the average nuclear delay had grown to 59 months.[48] The nuclear industry slipped so far from the puffed-up projections of the early 1970s

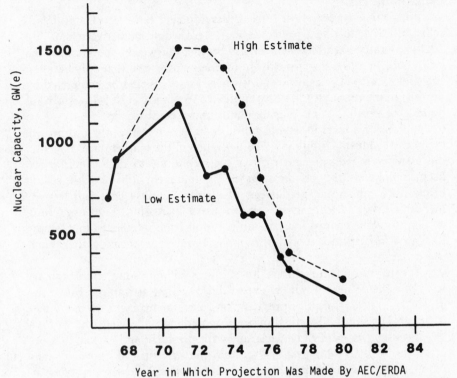

Nuclear Capacity Projected for U.S. in Year 2000

High Estimate

Low Estimate

Nuclear Capacity, GW(e)

1500

1000

500

68 70 72 74 76 78 80 82 84

Year in Which Projection Was Made By AEC/ERDA

GRAPHIC BY JOHN HEYMANN

FIGURE 15 Tumbling forecasts for nuclear reactor growth. AEC, ERDA, and DOE forecasts for the minimum and maximum expected number of nuclear plants in the year 2000 have fallen drastically since 1974. The dream of a large nuclear power program has been crushed under the rising cost of nuclear plant construction.

Sources: Nucleus, Union of Concerned Scientists, Vol. 3, No. 1, Fall–Winter, 1980.

U.S., Atomic Energy Commission, Civilian Nuclear Power—The 1967 Supplement to the 1962 Report to the President, Washington, D.C., 1967.

U.S., Atomic Energy Commission, Office of Industry Relations, The Nuclear Industry 1973, WASH-1174-73, Washington, D.C., 1973.

U.S., Atomic Energy Commission, Office of Industry Relations, The Nuclear Industry 1974, WASH-1174-74, Washington, D.C., 1974.

U.S., Atomic Energy Commission, Office of Planning and Analysis, Nuclear Power Growth 1974–2000, WASH-1139(74), Washington, D.C., 1974.

U.S., Department of Energy, Energy Information Administration, Commercial Nuclear and Uranium Market Forecasts for the United States and the World Outside Communist Areas, DOE/EIA-0184/24, Washington, D.C., 1980.

U.S., Senate, Subcommittee on Energy Research and Development of the Committee on Energy and Natural Resources, ERDA Fiscal Year 1978 Authorization, 95th Congress, lst sess., 24 March 1977, publication no. 95-80, Part I.

that in 1980, wood surpassed nuclear-generated electricity as an energy source in the U.S.[49]

Long-range forecasts for nuclear development have also fallen drastically. In 1974, the AEC predicted that there would be between 850 and 1,400 1,000-megawatt light-water and breeder reactors in operation by the year 2000. In 1981, the Department of Energy predicted that only between 155 and 195 1,000-megawatt light-water reactors would be completed by the turn of the century.[50] The low figure of 155 plants is just below the total number of existing plants plus all plants under construction.

Billions of dollars in federal subsidies have not been enough to protect the nuclear industry from the harsh discipline of the marketplace. In 1982, the industry is reeling: construction costs continue to rise; cancellations keep coming in; utility bond ratings have plummeted; utility stock is selling below book value; and predictions of dozens of utility bankruptcies—or nuclear Chrysler corporations—appear with increasing frequency.[51] According to some energy analysts, only a multibillion-dollar bailout by the Reagan Administration would be capable of preventing the industry's financial collapse.

The reactor manufacturers have suggested that some extraordinary steps be taken to resume nuclear expansion. Speaking at Stanford University in 1980, J. J. Taylor, vice-president and general manager of the Water Reactors Division at Westinghouse, proposed the following: that a "*national energy emergency*" be declared by the president and backed by Congress; that "all *regulatory delays*" be "*eliminated*"; that all "existing safety requirements" be met, but "*paperwork requirements,* such as quality assurance audit procedures" be "*streamlined*"; that "new requirements, such as those stemming from the *TMI lessons learned*" be met, but "not permitted to delay plant operation"; that financing be made available; and that plant modifications and improvements "continue to be made, but on an *orderly schedule.*" Taylor claimed that the industry was "ready for a *dramatic expansion,*" but "no industry can afford to maintain idle capacity for a market that does not exist."[52]

Nuclear developers paint a grim picture of what could happen if we do not build hundreds of reactors by the year 2000. Milton R. Copulos of the Heritage Foundation, an important source of Reagan Administration policy advisors, claims that halting nuclear expansion

> could result in significant *economic and social dislocation.* A shortage of over eight million jobs at an annual cost of more than $98 billion could develop. Blacks, women, and other groups just beginning to make economic headway would see recent gains disappear. Most importantly, attempts to cope with *large-scale shortages* of electrical generation capacity could result in *profound changes* in the American lifestyle.[53]

Is there no choice but to bail out the nuclear industry?

CHAPTER NINETEEN

Alternative Visions

The energy problem cannot be solved by a single technological panacea like nuclear power. It can be solved through the application of a host of measures, some technological and some institutional, designed to improve the efficiency with which we use energy and to shift our economy to solar sources.

Nuclear developers have been slow to abandon their misplaced hopes. Even as the nuclear industry began to collapse under the pressure of the marketplace, they continued to claim that a massive reactor program was necessary. Since the 1973 OPEC oil embargo, they have frequently presented nuclear power as the ticket to *energy independence*, ignoring—and sometimes actively opposing—more viable energy strategies. The *energy crisis* is not the result of a failure of technology, but of a failure of human vision.

Contrary to the beliefs that make up the nuclear mindset, nuclear power is incapable of making an important contribution to solving the energy problem. First, the cost of nuclear electricity is too high to permit continual growth in energy demand. Even with a lavish nuclear program, rapid exponential energy growth cannot be sustained. We would have to build new 1,000-megawatt reactors at an average rate of one every five days in order to meet 25% of the DOE's 1978 median forecast for energy use in the year 2000 with nuclear power.[1] Such a construction program would cost several hundred billion dollars a year, thus draining an enormous amount of capital from the economy. Continued exponential energy growth is simply too expensive.

Second, nuclear power alone cannot end dependence on foreign energy sources. The French nuclear program, for example, has been widely hailed by American nuclear developers as an example of a serious national effort to reduce dependence on foreign oil. A closer look at the French program shows the futility of trying to sustain exponential energy growth with nuclear power. One French projection called for the construction of the equivalent of 100 1000-megawatt reactors by the year 2000. This forecast predicted an energy growth rate of 3.8% per year (somewhat lower than the historical growth rate of 5.3% per year for the period 1963 to 1973). Even with the 100 new reactors, France would have to import 70% more oil and gas in the year 2000 than it did in 1975.[2]

Third, nuclear energy only produces electricity, which is not the crux of the energy problem. Most energy is used for heat and transportation, functions for which electricity is not particularly well suited. In 1952, Palmer Putnam believed that nuclear power would be cheap enough to justify the electrification of these functions. But his assumptions about the cost of nuclear power proved to be incorrect. Nuclear power is too expensive to justify widespread electrification of space heating and transportation.

In the mid-1970s some energy analysts recognized that the rising cost of nuclear construction and the collapse of utility growth forecasts had created a need for a new look at energy policy. A new energy strategy, which we will call the energy-efficiency approach, was developed. This strategy became the focus of international debate in 1976, when physicist Amory Lovins published an article called "Energy Strategy: The Road Not Taken?" in the journal *Foreign Affairs*. In the article, Lovins synthesized the work of dozens of researchers into a critique of current energy policy and an alternative strategy for meeting people's energy needs. Lovins argued that it would be possible to move the U.S. to a solar-powered economy in 25 to 50 years. He proposed a two-pronged strategy for accomplishing this transition. One prong was the use of technologies for improving the efficiency with which energy is used, thus curbing exponential energy growth and making the energy problem more manageable. The other prong was the gradual substitution of solar technologies for oil and gas.[3]

The energy-efficiency approach proposed by Lovins and his colleagues differs from traditional energy policy in several ways. The traditional approach emphasizes *energy production*, conceiving of the energy problem as one of increasing energy supply by finding ways to *produce* more energy. The energy-efficiency approach recognizes that it is possible to "produce" energy by improving the efficiency with which it is used. If the efficiency of oil use is doubled, the effect is the same as if world oil reserves were doubled.

The energy-efficiency approach relies on the principle of end-use analysis. Instead of simplistically viewing world or national energy use as a homogeneous lump of *energy demand*, energy-efficiency advocates look at the end-uses, such as heating homes or providing people with transportation, to which energy is put. They then ask what is the most efficient way to meet each of these needs. People do not want kilowatt-hours or British thermal units or barrels of oil; they want the services that energy provides, such as comfortably heated homes and businesses, reliable transportation, the convenience of refrigeration, and manufactured goods. Insulating houses turns out to be much cheaper than building nuclear power plants to run electric heaters. Making cars that get better gas mileage is cheaper than producing synthetic gasoline.

Traditional energy policy holds that growth in energy use is essential to produce growth in the gross national product (GNP). The energy-efficiency approach recognizes that a low-energy future need not be achieved by Draconian cutbacks in the standard of living or a return to a pastoral existence. Improvements in energy efficiency can lead to increased GNP without increased energy use. A 1981 study by the Solar Energy Research Institute (SERI), a government-supported laboratory, concluded that "far from being incompatible with vigorous economic growth," a strategy built around energy efficiency and solar sources "may actually be essential for such growth." Failing to take efficiency measures would have a detrimental effect on the economy, since rising energy prices would drive both inflation and recession.[4] SERI reported that the strategy it recommends could virtually eliminate oil imports in two decades, while the GNP increases 80%, and the population increases 17%.[5] (The U.S. imported about 6.8 million barrels of petroleum per day in 1980.)

The infrastructure of the U.S. economy was built during an era of cheap energy, and people expected energy to continue to grow cheaper. Little attention was paid to doing things efficiently. Today, the replacement cost of energy is soaring; each new barrel of oil and each new kilowatt-hour of electricity costs more than the last one did. As a result, there is tremendous potential for making improvements in energy efficiency, and most of these improvements rely entirely on existing technologies. In most cases it is cheaper to save a barrel of oil than to produce one. The Harvard Business School's Energy Policy Project concluded in 1979 that with a "serious commitment to conservation," the United States could use "30 to 40 percent less energy than it now does, and still enjoy the same or an even higher standard of living. That saving would not hinge on a major technological breakthrough, and it would require only modest adjustments in the way people live."[6]

Some nuclear developers have claimed that a future based on the energy-efficiency approach would be a *low-technology future*. A special report to the National Research Council in 1978 concluded that "probably just the reverse is true. A low-energy future offers strong incentives for technological innovation. In many recent inventions, information has been substituted for energy."[7] Consider, for example, the hand-held calculator. Twenty years ago, the same amount of computing power would have filled a large room with tubes and transistors and would have required large amounts of energy. This computing power now fits into a shirt pocket or onto a wristwatch and runs off tiny batteries. Improvements in energy efficiency made by substituting brain power for raw energy can eliminate the need for great increases in energy use.

Simple measures like insulating buildings can produce huge energy savings. Most homes are underinsulated in comparison to the level of insulation that is economically justifiable, and most new buildings are being built less weather-tight than they might be. The American Institute of Architects estimated in 1977 that the U.S. could save the energy equivalent of 12½ million barrels of oil per day by 1990 (nearly twice what the country currently imports per day) if new and existing buildings were designed to be energy efficient.[8] In 1980, Princeton researchers Robert Williams and Marc Ross called for the creation of a cadre of "house doctors" trained to locate the obscure air leaks often overlooked when insulating a house. They concluded that a national house-doctor program could

> "drill" the giant "oil and gas fields" represented by existing residential buildings and thereby "recover" the oil and gas now wasted. A successful effort in this direction could, by the late 1980s, save half the energy now consumed for space heating. From oil and gas-heated residences alone, this would "produce" the equivalent of 1.6 million barrels of oil per day, or about half our dependence on Middle Eastern and North African oil.[9]

Improving automobile gas mileage is another way to produce large energy savings. One out of every nine barrels of oil used in the world each day is burned as gasoline on American highways.[10] In 1978, the average car on the road in the U.S. got about 14 miles per gallon.[11] Replacing these cars with automobiles that get better gas mileage would save millions of barrels of oil per day. Chrysler's K Cars average 25 miles per gallon. The Volkswagen diesel Rabbit gets 42 miles per gallon. VW has developed a prototype turbocharged diesel Rabbit that has attained mileage of 78 miles per gallon.[12] The average vehicle is rarely in use for more than 10 to 12 years. Over the course of a decade, the average gas mileage of all the cars on the road in the U.S. could be raised considerably. The Solar Energy Research Institute concluded that it should be possible to increase the fuel efficiency of the U.S. auto fleet substantially without major sacrifices in comfort, safety, or performance: "If the average car on the road in the year 2000 obtains 55 mpg, the nation would consume nearly 3 MBD [millions of barrels per day] less gasoline than it does today, even with significant increases in both the population and the miles driven by each person."[13]

Numerous opportunities for major energy-efficiency improvements exist in industry. Process steam accounts for about 14% of all the energy used in the U.S.[14] In producing this steam, energy is wasted in the form of waste heat from boilers. Through a process called cogeneration, which is used widely in Europe, the waste heat could be used to produce electricity. End-use analysis shows that cogeneration provides a much more efficient way to make electricity and produce steam than does producing the same quantities of each separately. The amount of fuel needed to produce a kilowatt-hour of electricity by cogeneration, beyond what is needed to produce the steam, is only about one-half of what is needed to produce a kilowatt-hour of electricity at a central-station power plant. If all large industrial boilers were involved in cogeneration, the electricity produced would be the equivalent of that produced by 30 to 200 nuclear plants, depending on the technology used.[15] In addition to cogeneration, many other improvements in industrial processes could be made that would cut energy use drastically.

The combination of energy-efficiency measures in the areas of space heating, automobiles, and industry could save considerably more oil than we now import. The measures involved use already proved technologies. Moreover, these steps can be taken quickly, leading to a reduction in oil imports much more rapid than any reduction that could result from nuclear construction.

Though the nation's official energy policy still holds that *energy production* is the way out of the crisis, the efficiency approach is already beginning to be implemented successfully. In 1979, energy analyst Vince Taylor found that improvements in energy efficiency had far outpaced increases in energy production during the years 1973 to 1978. Taylor found that during this period, efficiency improvements provided two and one-half times as much energy as did increases in energy production.[16]

Energy-efficiency measures alone, of course, are not sufficient to supply the energy services people want; even energy that is used very efficiently must be supplied by some means. But energy-efficiency measures are crucial to a workable energy policy. Not only do they save money, but they also extend the life of fossil-fuel reserves and buy time to develop other energy sources. Most important, energy efficiency transforms the energy problem from one of trying to supply ever rising quantities of energy to one of finding ways to supply a fixed amount of energy after exponential energy growth stops and energy use reaches a plateau. Curbing exponential energy growth would make it unnecessary to divert exponentially growing quantities of capital into energy production. It would also greatly reduce environmental and safety hazards, since it would make it unnecessary to develop a large breeder program or to mine and burn ever rising quantities of coal.

Proponents of the energy-efficiency approach see reducing energy growth as the first step toward solving the energy problem. The second step is to gradually substitute solar energy sources for fossil fuels. There are a wide range of technologies for deriving energy from the sun. These include solar thermal energy (for heating and cooling buildings, or for providing hot water for agricultural or industrial process heat), fuels from biomass (e.g., wood and alcohol), and solar electric energy (e.g., photovoltaics, solar thermal electric, wind power, and hydropower). This diversity of technologies means that solar's potential contribution is not dependent on the success of any particular technology, since difficulties in one area would not impede success in another.

A number of different studies have concluded that solar can make an important contribution to U.S. energy needs by the turn of the century. In 1978, the president's Council on Environmental Quality (CEQ) predicted that solar energy could account for 25% of the country's energy by the year

2000.[17] In the fall of 1978, a special 30-federal-agency review of the potential for solar energy estimated that solar energy might contribute as much as 20% by the year 2000.[18] In June 1979, President Carter announced a national goal of obtaining 20% of the country's energy from solar by the year 2000.[19] In a study published in 1980, the Union of Concerned Scientists concluded that a transition to a high-technology, all-solar economy could be largely accomplished by the year 2050.[20] And a DOE-sponsored study of the energy future of the state of California concluded that solar-based technologies could meet almost all of the state's energy needs by the year 2025.[21]

A number of solar technologies are already available. Solar heating is already a viable alternative to imported oil, according to the Harvard Business School Energy Project.[22] The production of alcohol fuels from biomass has taken off rapidly. In Brazil, a long-range plan to convert the entire automobile sector to pure alcohol is under way. The major world auto manufacturers are planning to churn out hundreds of thousands of cars in Brazil which burn 100% alcohol.[23] Wind power is making a comeback, both for generating electricity and for powering ocean-going ships. In 1980, *Science* reported that "wind power, which only a decade ago was greeted with bemused and tolerant smiles, is now attracting the interest and enthusiasm of the nation's hard-nosed utility officials." Utilities investing in wind-generating equipment include Southern California Edison, the Nova Scotia Power Company, and the Hawaiian Electric Company.[24] Ships consume 5% of the world's oil, and substantial savings can be made by developing high-technology sailing ships.[25] Japan has already launched a small, 1,600-ton tanker equipped with folding plastic sails controlled by a computer. And a large Japanese shipbuilding firm is constructing an 80-foot test model of a 300,000-ton wind-powered supertanker.[26] Research and development of photovoltaics—which convert sunlight directly into electricity—continue to produce promising results. Photovoltaics were originally developed for powering space satellites, and the cost per watt was very high. But the cost has been dropping steadily since the mid-1960s, and many analysts expect it to continue to fall as the manufacturing process is automated.

No grand *technological breakthroughs* are necessary to begin implementing a transition to an economy based on the efficient use of renewable solar resources. Energy-efficiency and solar technologies are already available at cost-effective prices. But a number of what Lovins terms "institutional barriers" are inhibiting investment in the energy-efficiency approach.[27] These institutional barriers include tax laws, building codes, utility hostility, and inadequate capital access.

As an example of how these institutional barriers interact to slow investments that would be in the national interest, consider the case of cogeneration. As we have said, cogeneration is a cost-effective way of producing electricity compared to nuclear power, and it is an important alternative for displacing new nuclear capacity. But few companies are taking advantage of this potential because utilities have long been hostile to cogeneration and have established rate-setting practices to discourage its use.

In 1920, 22% of the electricity in the U.S. was generated by cogeneration at industrial sites, but by 1976, the percentage had fallen to only 4%. Utilities discriminated against cogeneration in two ways: they charged very high backup rates for supplying electricity when a cogenerating system was out of operation. And they refused to buy excess electricity from companies that cogenerated except at very low prices.[28]

Similar institutional barriers are impeding the development of many energy-efficiency and solar technologies. Let us repeat that none of these problems are technological. All of them can be solved by the right combination of laws, regulatory changes, and tax incentives, provided that the political will exists to make these changes. The belief that "we have *no choice*" is a self-fulfilling prophecy.

A reexamination of the energy problem is beginning all over the country. People are acting to remove the institutional barriers that are preventing the expansion of energy-efficiency and solar technologies. In December 1979, the California Public Utility Commission (PUC) imposed a $7.2 million penalty on Pacific Gas and Electric Company for failing to pursue cogeneration opportunities aggressively enough.[29]

In California in 1980, the PUC ordered the state's four largest publicly owned utilities to offer their customers low-interest loans and rebates to finance the installation of solar hot-water heating equipment. The program was designed to reach 15% of the homes in the state with electric hot-water heating, leading to the installation of 175,000 solar water heaters. The solar water heaters are expected to save the equivalent of one million barrels of oil per year.[30]

In Seattle, Seattle City Light, the country's third largest municipal utility, was asked to participate in the construction of two new nuclear plants. After a 14-month-long review, the utility rejected the nuclear plants in favor of a program emphasizing energy efficiency. The city recognized that energy efficiency was cheaper and enacted a series of measures to carry out the new program.[31]

The most important barrier to the development of a sound energy policy is the nuclear mindset, which persists within the DOE and most of the energy industry. Because of the nuclear mindset, many federal energy policy makers have not taken the energy-efficiency approach seriously. Energy-efficiency and solar technologies have been consistently underfunded. During the early 1950s, these technologies did not get a fair hearing. There were no technical barriers to pursuing further research on these technologies, and some could have moved into commercial use quickly. Cogeneration was widespread during the 1930s and 1940s; solar water heating was used extensively in the U.S. in the early 1900s. But the nuclear mindset, embodied in the policies of the AEC and the Joint Committee on Atomic Energy, stood in the way. The nation vested its hopes—and invested its money—in a nuclear future.

The AEC and the JCAE often showed contempt for solar power. This disdain was apparent during the 1972 JCAE hearing on the AEC's 1973 fiscal-year budget. The AEC had been given the authority to do non-nuclear energy research, and Senator Mike Gravel of Alaska submitted an amendment to the authorization bill requiring the AEC to spend $15 million on solar energy research. During a JCAE hearing, Representative Hosmer mockingly compared Gravel's proposal to someone who "tells you about filling up a gasoline tank full of salt water and dropping a pill in it and thereafter it runs to 100 miles per gallon. . . ."[32] Representative Holifield said that "with due respect to the learned gentleman from Alaska . . . this is a *fantastic proposal*, for us to be talking about spending money on sea thermal gradients and wind power and light, heat, and the process of photosynthesis when we are in the midst of a crisis which has to be solved immediately."[33]

The federal solar energy research budget at the time consisted almost entirely of a tiny $1 million program being run by the National Science Foundation. Representative Holifield asked the AEC's Milton Shaw, the head of the Division of Reactor Operation, whether he thought a poll of the scientific community would reveal support for the NSF program. Shaw replied, "Actually, I think for example, that a million dollars being spent this year by NSF for solar energy is really far greater than perhaps one could justify from taking a poll [of scientists]. . . ."[34]

In the latter part of the 1970s, federal support for solar energy increased rapidly, reaching about $600 million a year during the last year of the Carter Administration. (The solar budget was still less than half of the total nuclear budget for the same year, $1.43 billion.) The Reagan Administration has reverted to the mindset of the AEC toward solar power and conservation. In the spring of 1981, Reagan proposed slashing the solar budget for fiscal year 1982 by 67% compared to the previous year, and reducing the conservation budget by 75%. Reagan also proposed raising the nuclear fission budget by 16% and raising the total nuclear budget by 24%. Under Reagan's proposed fiscal-year 1982 budget, federal expenditures on nuclear power would be more than five times greater than all expenditures on solar and conservation combined.[35]

Nuclear energy supplies only about 3½% of the nation's energy, less than we now obtain from firewood.[36] But despite the failure of the nuclear industry, nuclear developers continue to promote nuclear construction as the answer to the energy crisis. An ad published by "America's Electric Energy Companies" in the *New York Times Magazine* in 1981, for example, raised the specter of a war fought over oil, suggesting that nuclear power was necessary to help avert such a war. The double-page ad pictured three young men (two white and one black) being sworn into the armed forces. The headline read: "Maybe The World Won't Go To *War Over Oil* . . . But Who'd Be Foolish Enough To Want To Find Out?" One subhead advocated "Replacing Imported Oil with Nuclear Power," and the ad recommended using more nuclear power to achieve "Energy Independence And World Peace."[37]

The mindset presented by ads like this one is contributing to the risk that the world will end up going to war over oil. Nuclear power can do little to reduce dependence on imported oil. It is too expensive, too slow to come on line, and does not address the critical problems of heating and transportation. Yet the nuclear *magic wand* has the country's energy planners spellbound. The illusion that nuclear power can play an important role in solving the energy crisis is slowing the development of an energy policy capable of ending oil imports. We cannot spend the same money twice: funds tied up in nuclear construction cannot be used to effect a transition to renewable sources. The longer we retain faith in the nuclear panacea, the more costly and difficult the transition will be, and the greater will be the risk that a war over oil will occur.

PART EIGHT

Atoms for War

Scheer: Don't we reach a point with these strategic weapons where we can wipe each other out so many times and no one wants to use them or is willing to use them, that it really doesn't matter whether we're 10% or 2% lower or higher?

Bush: Yes, if you believe there is no such thing as a winner in a nuclear exchange, that argument makes little sense. I don't believe that.

Scheer: How do you win in a nuclear exchange?

Bush: You have survivability of command and control, survivability of industrial potential, protection of a percentage of your citizens, and you have a capability that inflicts more damage on the opposition than it can inflict on you. That's the way you can have a winner, and the Soviets' planning is based on the ugly concept of a winner in a nuclear exchange.

Scheer: Do you mean like 5% would survive? Two percent?

Bush: More than that—if everybody fired everything he had, you'd have more than that survive.

Interview of presidential candidate
George Bush by Robert Scheer of
the Los Angeles Times *January,*
1980[1]

Strategic superiority translates into the ability to control a process of deliberate escalation in pursuit of acceptable terms for war termination. The United States would have a politically relevant measure of strategic superiority if it could escalate out of a gathering military disaster in Europe, reasonably confident that the Soviet Union would be unable or unwilling to match or overmatch the American escalation. It follows that the United States has a fundamental foreign policy requirement that its strategic nuclear forces provide credible limited first-strike options.

Colin S. Gray
Director of National Security
Studies, The Hudson Institute,
1978[2]

CHAPTER TWENTY

The Hall of Mirrors

The world of nuclear warfare is a world of doublethink, a hall of mirrors, where *peace* is preserved through the constant threat of war, *security* is obtained through mutual insecurity, and nuclear war planners "think about the *unthinkable*," holding millions of civilians hostage to the most powerful death machines in history.

Nuclear war strategists have developed an esoteric, highly specialized vocabulary. In their ultrarational world, they talk in cool, clinical language about *megatons* and *megadeaths*. Cities are *bargaining chips*; they are not destroyed, they are *taken out* with *clean, surgical strikes*—as if they were tumors.

Since there is no way to defend cities and industry against nuclear attack, *global stability* is now preserved through a system called the *balance of terror*. The balance of terror is based upon the principle of *deterrence*. Nuclear deterrence is, in effect, a mutual suicide pact: if you attack me, it may kill me, but I will kill you before I die. The civilian population of each superpower is held hostage by the opposite power. The same is true of the allies covered by the *nuclear umbrellas* of the superpowers. If either side attacks, all the hostages will be destroyed.

The U.S. Department of *Defense* (known as the War Department until 1948) is incapable of defending the United States against a nuclear attack by the USSR. But it is capable of killing many millions of Russian citizens at the push of a button, and the Russians are incapable of doing anything to prevent the carnage. If the Russians were to attack the U.S. or its allies with nuclear weapons, the U.S. would retaliate by attacking the USSR. Since the Russians know this, the reasoning goes, they will be deterred from striking first. A mirror-image argument describes how Russia deters the U.S. from striking its people.

Military analysts classify nuclear attacks as either *counterforce* or *countervalue* attacks. A counterforce attack is one that is directed primarily against the other side's military forces; countervalue attacks are directed against cities and industry.

That a counterforce attack is directed primarily against military forces does not mean that there would be few civilian casualties. On the contrary, counterforce attacks could leave millions of civilians dead from the fallout produced by attacks on missile silos. A large airport might be construed as a military target because it could serve as a base for military planes. A naval base situated in a metropolitan area is another example of a target that could be interpreted as either counterforce or countervalue. The difference between a counterforce and a countervalue attack is not whether civilians die, but whether this is the main goal or a side effect. The deaths of civilians and the destruction of nonmilitary property in a counterforce attack is called *collateral damage*.

Nuclear war planners have always been afraid that the other side might try to launch a *preemptive first strike* (also known as a *splendid first strike*), that is, a counterforce attack designed to cripple the enemy's ability to retaliate. The reasoning behind a preemptive first-strike strategy goes as follows: If country X can destroy a large enough number of country Y's missiles in a first-strike attack, then X can also threaten to destroy Y's remaining cities in a second *nuclear salvo* if Y retaliates with any remaining missiles. It is assumed that Y will be *rational* and surrender to X rather than ensure its total destruction by launching an attack for revenge. X can then impose its will on Y, thus *winning* the nuclear war.

In its public statements, the USSR has renounced the *first use* of nuclear weapons; the United States has not. Nevertheless, military planners in both countries prepare for *worst-case* situations and tend not to believe verbal declarations. In war, the argument goes, *capabilities* count more than *intentions*, since intentions change without warning.

As a result of this perception of the possibility of an enemy attempting a splendid first strike, the two superpowers have engaged in a massive *arms race*, reaching higher and higher levels of destructive power. Some years ago, each superpower attained *overkill*, the ability to kill every citizen on the other side more than once. Nevertheless, the arms race continues, and each side continues to expand and *modernize* its nuclear arsenal.

The driving force behind the arms race is a treacherous double-bind known as the *security dilemma*. Since X is afraid of Y's weapons, X adds to its arsenal. X's arms buildup, which is conceived of as *defensive* by X, is perceived as *offensive* by Y, prompting Y to build more weapons to deter an attack by X. X looks at this and concludes that Y must be planning to attack; otherwise Y would not have expanded its arsenal. X therefore decides to build still more weapons, and the cycle continues.

The security dilemma has led to the creation of huge military establishments in both the U.S. and the USSR. The superpowers' *hawks* watch each other closely, passing what they see through the gloomy filter of worst-case *scenarios*. The hawks of one nation contribute to the prestige and power of the hawks of the other, and arms budgets climb.

The arms race has produced a wide array of nuclear weapons and delivery systems for getting them to their targets. The weapons are designed for use in different situations and vary considerably in explosive power. Nuclear warriors generally divide these weapons into three categories: strategic, tactical and theater.

Strategic nuclear weapons have high-yield warheads; each warhead may be hundreds of times as powerful as the bomb that destroyed Hiroshima. The Hiroshima bomb had a *yield* of 13,000 tons of TNT, or 13 *kilotons*;[3] strategic weapons often have a yield measured in *megatons*— millions of tons of TNT. Strategic weapons are capable of striking targets many thousands of miles away. Both superpowers have deployed their strategic weapons in bombers, in land-based missiles, and in submarines. In the U.S., this three-legged war machine is called the *strategic TRIAD*. Bombers armed with nuclear weapons wait for the *go code*. Land-based *Intercontinental Ballistic Missiles (ICBMs)* are ready to strike at the push of a button. Submarines bearing nuclear-armed missiles are *on-station*, waiting for a transmission that would order them to launch their cargo. Radar systems scan the sky for incoming missiles or bombers. At least one of the *Strategic Air Command's* flying command posts, officially called the *Looking Glass Planes* and unofficially known as the *Doomsday Planes*, is in the air at all times. And overhead, a network of satellites circles the earth, watching Soviet ICBM fields and maintaining *command, control and communications*, or C^3 (*C cubed*)—the capability to transmit and receive information and orders.

Tactical nuclear weapons are designed for use on the *nuclear battlefield*. They have much smaller yields than strategic weapons, their yields usually range from as low as one kiloton to several times the yield of the Hiroshima bomb. Tactical nuclear weapons are designed for use in conjunction with *conventional military forces*. In a land war, tactical nuclear weapons might be used to *take out* enemy tank columns. On the sea, they could be used to sink enemy warships. They can be shot from artillery, dropped from planes, shot in short-range missiles, used in depth charges or torpedoes, or placed in land mines.

Theater nuclear weapons have powerful warheads like those of strategic weapons. They do not have intercontinental range, however, and are designed for use in a *limited theater of operations*—like Europe, for example. These weapons include bombers and missiles with medium to long ranges. They are *deployed* on land and on aircraft carriers.

The U.S. government has also developed a new kind of nuclear weapon called the *neutron bomb* or *enhanced radiation warhead*. Neutron bombs produce less explosive blast than other nuclear weapons, releasing a greater fraction of their energy in a deadly burst of neutron radiation. Neutron bombs purportedly make it possible to kill enemy troops while reducing blast damage to the surrounding countryside. In its war games, the Defense

Department envisions using neutron bombs to stop Soviet tank attacks in western Europe.

Equipped with this array of armaments, nuclear warriors are ready to play the game of *escalation*, using threats and counterthreats to deter, influence, coerce, and block their opponents.

The theory of *limited war*—war which the combatant nations limit in scope or intensity by tacit or explicit agreement—is important to escalation strategy. This theory holds that war can be limited by restricting the geographic region in which it is conducted, by limiting the kinds of weapons used, or by limiting the kinds of targets attacked.

Nuclear war strategist Herman Kahn outlined a theory of escalation and limited war in a 1965 book called *On Escalation: Metaphors and Scenarios*. Kahn developed an *escalation ladder* with forty-four rungs, or levels of conflict. The rungs Kahn described range from "Political, Economic and Diplomatic *Gestures*" through "*Nuclear 'Ultimatums'* " and limited evacuation of cities, before crossing the "*No Nuclear Use Threshold*." From "*Local Nuclear War*," the ladder rises to "*Exemplary Attacks*" on property and population, before reaching "*Slow-Motion Countercity War*." As the intensity of the conflict climbs, the level of "*Countervalue Salvo*" is reached, and finally, the orgasmic release of "*Spasm or Insensate War*,"[4] as everyone lets loose with everything they have got.

Escalation strategy is a complex game of *nuclear chicken*. Opposing strategists, like two drivers headed on a collision course, try to force each other to back down by threatening terrible consequences for both unless somebody backs down. A disagreement might escalate into a crisis, a crisis into a conventional war. The use of tactical nuclear weapons would escalate conventional war into *limited nuclear war*. If this happens, no one knows whether the use of nuclear force could be neatly contained. Some analysts fear that crossing the *no-nuclear-use threshold* would ultimately lead to a *spasm war*.

Escalation strategy, also known as *brinksmanship*, is ripe with paradox. Survival depends on everyone being *rational*, yet it is hard to tell what the word rational means. Sometimes it seems rational to pretend to be irrational, even to act irrationally, making the illusion more credible by making it more real. In a game of chicken, the driver who throws his steering wheel out the window has won control of the road. Similarly, the nuclear warrior can seize the advantage by throwing away options, or by convincing the opponent he is willing to plunge over the brink.[5] President Nixon, for example, developed a strategy he called the "*Madman Theory*" to try to force the North Vietnamese to negotiate. According to Nixon operative H. R. Haldeman, convicted in the Watergate coverup, Nixon said:

> I want the North Vietnamese to believe I've reached the point where I might do *anything* [original emphasis] to stop the war. We'll just slip the

word to them that "for God's sake, you know Nixon is obsessed about Communism. We can't restrain him when he's angry—and he has his hand on the *nuclear button*"—and Ho Chi Minh himself will be in Paris in two days begging for peace.[6]

Over the years, the U.S. government has developed a number of theories about how to maintain deterrence; these are known as *strategic doctrines*. The best known is the strategy of *mutually assured destruction (MAD)*, developed by Robert McNamara, Secretary of Defense during the Kennedy and Johnson administrations. Under this strategy, nuclear war is deterred by the threat that any attack would promptly lead to a *nuclear exchange* that would destroy both superpowers.

For the balance of terror to remain *stable*, nuclear war strategists must keep escalation under control. Each side must believe that everyone's nuclear forces have *survivability*, the ability to survive a counterforce attack and still deliver a crippling retaliatory blow. Both sides must believe that the *costs of striking* would be greater than the *costs of not striking*.

If the survivability of either side's forces is in question, the whole situation becomes a hall of mirrors. What if X thinks Y thinks X could take out Y's weapons in a preemptive first strike? Should X strike? If X doesn't, Y might strike first, because X thinks Y might think it has nothing to lose. And what might Y think about all this? A terrifying web of perceptions and misperceptions is possible. *Spiraling tensions* could start a thermonuclear war even if no one wanted it.

During the late 1960s and the early 1970s, the survivability of each superpower's nuclear forces was not in question. In the past decade, however, improvements in weapons technology have made the survivability of land-based missiles less certain. This erosion of survivability is the result of what is known as *technological creep*, improvements in weapons technology that seem to have a momentum of their own.

One of the most *destabilizing* technological developments of the 1970s was the deployment of *MIRVs, multiple independently targetable reentry vehicles*. MIRVs make it possible for a single missile to carry a number of nuclear warheads, each of which can be aimed at a separate target. The U.S. began deploying MIRVs in 1970, and the Soviet Union began in 1975.[7]

MIRVs tend to give the advantage to the side which strikes first in a nuclear exchange. A quick look at the following example will illustrate why this is so. Imagine a situation in which each side has 1,000 missiles with 10 MIRVed warheads on each missile. By striking first with 100 missiles—MIRVed with 1,000 warheads—the attacker could eliminate all of the other side's missiles. This would leave the attacker 900 missiles to use as a deterrent against retaliation. While this example is hypothetical, the message is clear: MIRVs are destabilizing.

A second case of technological creep has occurred in the area of missile accuracy. Extreme accuracy is not important for *city-busting*, since the target is large and *soft*—unprotected and easily destroyed. Accuracy is important for counterforce attacks, however. Underground missile silos, with their heavy shieldings, are very *hard* targets, and to destroy them, it is necessary to make a *direct hit*. Over the past decade, each superpower has greatly improved the accuracy of its missiles, so much so that they can now land within a few hundred feet of their targets.[8] Weapons specialists refer to this increase in accuracy as a decrease in the *CEP*, or the *circular error probable*, which is the radius of the circle in which a missile has a 50% chance of landing if aimed at its center. As a result of the increase in accuracy, land-based missiles in both the U.S. and the USSR are vulnerable to counterforce attack.

No one has developed *antisubmarine warfare (ASW)* technology capable of threatening the survivability of either U.S. or USSR submarines, and the subs remain a *credible* deterrent. Work to *improve* antisubmarine warfare is under way in both countries, however.

The C³ systems both superpowers depend on to coordinate their nuclear forces might be vulnerable to nuclear attack. This could provide a *strong incentive* to strike first. As the newsletter of the Federation of American Scientists (FAS) noted in October 1980,

> A nation that strikes first with strategic forces does so with its command structure, control mechanisms, and communications devices wholly intact, alerted, and ready. Each and every telephone line, satellite, and antenna is functioning and every relevant person is alive and well. By contrast, the nation which seeks to launch a retaliatory attack may find its chain of command highly disrupted, its telephone lines dead, its satellites inoperative, its radio signals interfered with, and its communications officers out of action.[9]

The FAS called attacks on C³ "a kind of supercounterforce and correspondingly destabilizing." The Federation predicted: "Should either side carry out deliberate efforts to attack the C³ of the other, it appears almost certain that a spasm war would result in which the attacked nation gave its military commanders either by prior agreement or by last desperate message, the authority to *fire at will*. As its ability to communicate gave out, it could and would do no less than use its last communications channel for *the final order*."[10]

Another threat to the stability of deterrence is the possibility that a system failure in either superpower's nuclear-war-fighting computers could trigger an accidental nuclear war. Three recent *alerts* caused by computer errors show that this threat may not be as insignificant as the Department of Defense claims:

☐ In November of 1979, data from a computer *war game* accidentally flowed into a live warning and command network, triggering a low-level alert. The computer's mistake was not detected for six minutes. In the meantime, B-52 pilots were told to man their planes, and the launch officers in ICBM silos unlocked a special strong box, removed the *attack verification codes*, and inserted the *keys* into their slots. When two keys ten feet apart are turned within two seconds of each other, the missiles blast off.[11]

☐ In June, 1980, on two separate occasions, a computer error caused by a faulty circuit chip worth 46¢ sent out false signals that the USSR had launched missiles headed for the U.S. In both cases, some of the B-52 fleet started its engines before the error was detected.[12]

The Pentagon maintains that there is *"no chance that any irretrievable actions* would be taken on the basis of *ambiguous computer information*,"[13] noting that the computers do not make decisions alone and that *human intervention* has always detected the errors.

Nevertheless, there is little time for the people involved to read the signals properly and make decisions. Land-based missiles can reach the U.S. in about thirty minutes, while submarine-launched missiles might take only half that time. Moreover, the threat exists that an erroneous alert could generate a nuclear attack as if by a trick of mirrors. If in response to a computer error the B-52s were suddenly to take off from their air bases, the Russians would immediately detect the maneuver. Soviet officers would have even less time to reach a judgment about how to respond, since their *early warning systems* are not as sophisticated as those of the U.S. If the Soviets dispatched their bombers, the U.S. warning system would in turn detect the planes, and the computer's message, though originally erroneous, would be *confirmed*.

The Pentagon claims that such a *chain reaction* is a *"highly unlikely scenario*."[14] A full public discussion of the issues involved is impossible because most of the relevant information is classified.[15]

CHAPTER TWENTY-ONE

The Shatterer of Worlds

Nuclear weapons release quantities of energy so far beyond the realm of ordinary experience that the effects of even a single nuclear blast are difficult to comprehend. Terms like *kiloton* and *megaton* are hard to grasp. Most people think of them as meaning something like "a hell of a lot." The effects of full-scale nuclear war can barely be imagined.

A short book called *A-Bomb: A City Tells Its Story* by the Hiroshima Peace Culture Center, which was founded by the city in 1967, gives a powerful description of the human suffering produced by a single 13-kiloton weapon. The book often quotes survivors of the 1945 Hiroshima bombing by the U.S., which killed more than 118,000 people. (The atomic bombing of Nagasaki killed more than 73,000 people.)[1] The book tells how the attack began:

> an eye-stabbing flash, a searing heat blast and a heaven-splitting roar . . .
> No one knew what had happened, only that people turned to charcoal here
> and there . . . or were tossed through the air . . . or crushed under falling
> buildings. Limbless or headless bodies rolled about or piled up like logs on
> the ground, and around and between them writhed the still living, their
> flesh torn and tattered.[2]

Rescue teams quickly began trying to help the wounded. They were hindered by the rapidly spreading fire. Medical facilities were hopelessly inadequate. "Eighteen hospital and 32 first-aid centers were destroyed or rendered useless for a time. 90% of the doctors were either killed or injured, and most of the nurses died."[3] A large number of patients had serious burns. They "were covered with serum from broken blisters which in many cases glued the clothing to their bodies." They received at best makeshift treatment. Sesame oil was applied to burns, since the supply of better medicine was limited. Patients were treated without anesthesia. People with burns on more than a third of their bodies were considered hopeless and simply abandoned. Wounds festered in the summer heat. "Maggots swarmed on living bodies, something never seen before."[4]

Disposal of the bodies was a major undertaking:

Countless people seeking water had crept to the river bank and died. Their chilled bodies became ice pillows for others, keeping them barely alive. Almost naked bodies floated by the hundreds in the river. . . . Volunteer wardens would touch a hand only to have the skin come away; there was no firm place to grasp. Of necessity abandoning human dignity, they finally used gaff hooks to lift them onto trucks. . . . [Within two weeks] the military and police together had disposed of 32,959 dead, but no one knew how many had floated out to sea.[5]

Many people died of radiation sickness, suffering terrible agony: hemorrhaging in the stomach; failure of such organs as the lungs, kidneys, spleen, and liver; vomiting, sometimes of blood; bloody excrement; and finally death. Doctors who performed autopsies on people who had died without visible wounds found that the "stomachs looked as though they had been curried with a wire brush . . ."[6]

Since Hiroshima, nuclear weapons have become much more powerful. In March, 1954—three months after Eisenhower's *Atoms for Peace* speech—the AEC tested a 15-megaton thermonuclear *device*, more than 1,000 times as powerful as the weapon that destroyed Hiroshima.[7] The test, known as the *BRAVO shot*, spread lethal levels of fallout over 7,000 square miles of the Pacific Ocean.[8] Radioactive fallout contaminated the crew of a Japanese fishing vessel, some Marshallese natives, and a group of American servicemen, creating an international incident (see chapter 7).

On March 31, AEC Chairman Lewis Strauss and President Eisenhower held a news conference on the BRAVO shot. When the press asked for a general description of what happened when a hydrogen bomb went off, Eisenhower announced that the information was classified. But the press did manage to badger Strauss into giving an indication of the power of the new weapon:

Strauss: . . . an H-bomb can be made large enough to *take out a city.*
Chorus: What?
Mr. Smith: How big a city?
Strauss: Any city.
Question: Any city—New York?
Strauss: The metropolitan area, yes.[9]

That this announcement shocked the Washington press corps is not surprising considering the efforts of Truman's and Eisenhower's adminstrations to restrict discussion of the hydrogen bomb. Under Truman, the AEC censored Hans Bethe's article on the hydrogen bomb (see chapter 5). And according to a diary kept by AEC Chairman Gordon Dean, on May 27, 1953, Eisenhower

made the suggestion that we leave "thermonuclear" out of press releases and speeches. Also "fusion" and "hydrogen" . . . . The President says *"keep them confused* as to 'fission' and 'fusion.' "[10]

In the summer of 1956, after a series of hydrogen-bomb tests in the Pacific sparked worldwide demonstrations, AEC Chairman Lewis Strauss tried to keep the public confused by saying that "the current series of tests has produced much of importance not only from a military point of view but from a *humanitarian aspect.*" Strauss insisted that *"real progress"* was being made in producing the "maximum effect in the immediate area of the target with *minimum widespread fallout hazards.*"[11] A storm of ridicule greeted what the press quickly labeled as *humanitarian bombs* and *clean bombs.* JCAE member Chet Holifield vigorously attacked Strauss's statement, saying that it would take "a miracle" for an atomic or hydrogen bomb to explode without producing "dangerous radioactive fall-out." Holifield said he was "concerned over the repeated pattern of applying advertising agency techniques to alleviate the natural fears of hydrogen destruction which trouble our people and the people of the world. Huckstering is no substitute for policy as we grope for solutions to the atomic and hydrogen dilemma."[12] Linus Pauling, one of the leading critics of the arms race and weapons testing, wrote that ". . . to call any weapon that can kill millions of people a *'clean'* bomb is to insult a noble word in the English language—the word *clean.*"[13]

In its 1957 hearings on fallout, the JCAE tried to pin down the AEC's use of the word *clean.* Representative Holifield began by questioning Dr. Alvin C. Graves, who had been the science director of Los Alamos since 1950, and the test director of the Nevada Test Site since 1951.

> Representative Holifield. No. This committee is not responsible for the phrase "clean bomb." We are not responsible for it. But there are millions of people throughout the world that may be hanging their hopes upon the fact that we have a humanitarian hydrogen bomb.
>
> Dr. Graves. I am afraid the only comment one can make on it is that *"cleanliness"* is a little bit relative anyway. What you mean by "cleanliness" in this case is a *question of degree.*
>
> Representative Holifield. You would not say in this case that cleanliness is next to godliness.
>
> Dr. Graves. No. . . .[14]

Later in the hearings, Holifield forced Graves to sum up his position on "clean" bombs:

> Representative Holifield. Therefore, the conclusion we can reach is that there is a dirty bomb and there is no such thing as a clean bomb, and I am using the word "clean" in the absolute sense and "dirty" in the absolute sense.
>
> Dr. Graves. There are *dirtier bombs*, and some that are *less dirty.*[15]

The committee then turned to AEC Commissioner Willard Libby, a more truculent witness. According to Libby:

> If you call cleaner "clean," maybe you are telling a *small fib*. But we certainly have a cleaner weapon, as you know. Now, the question of the adjective which is applied to that, all I can say is that perhaps we have used the *wrong adjective*, and perhaps we should say *"cleaner."*[16]

In keeping with its clean bomb nomenclature, the AEC measured fallout doses in *Sunshine Units (S.U.s)*. In 1953, the AEC began *Project Sunshine*, a secret program to monitor the accumulation in the environment and in humans of weapons-produced strontium-90, a radioactive bone-seeker like radium. The AEC invented the Sunshine Unit, which measured the ratio of strontium-90 to normal calcium. By 1957, the AEC had made Project Sunshine public. AEC officials denied that the use of the word *sunshine* was intended in any way to be misleading. Gordon Dunning of the AEC's Division of Biology and Medicine explained at the 1957 JCAE hearings on fallout that "the Sunshine Unit is a coined phrase. . . . Just like when one buys milk, you have to have some unit. It is merely a coined unit so that one in the business may know how much strontium 90 you are talking about. . . ."[17] AEC Commissioner Willard Libby told the JCAE that "perhaps we did not pay too much attention to the name in selecting it. But there was never any intent to mislead or to minimize the importance of the hazards."[18]

Choosing the right words to describe the effects of nuclear weapons has always presented a problem. In July, 1946, during *Operation Crossroads*, the first series of atomic-bomb tests covered by the media, observers witnessed a stunning display when the *Test Baker* atomic bomb was detonated underwater (see plate G). William Shurcliff, the official historian of Operation Crossroads, commented as follows:

> No adequate vocabulary existed. . . . The *vocabulary bottleneck* continued for months even among the scientific groups; finally, after two months of verbal groping, a conference was held and over thirty special terms, with carefully drawn definitions, were agreed on. Among these terms were the following: *dome, fillet, side jets, bright tracks, cauliflower cloud, fallout, air shock disk, water shock disk, base surge, water mount, uprush, aftercloud.*[19]

Since Operation Crossroads, nuclear weapons specialists have developed a full vocabulary to describe weapons effects. The energy from a nuclear explosion is released in a number of forms: some as explosive *blast*; some as radioactive fission products, forming *fallout*; some as *thermal radiation*, or heat; some as *direct radiation*; some as a massive pulse of electromagnetic waves, known as *EMP* for *electromagnetic pulse*.

PLATE G Twenty-three-kiloton underwater atomic bomb test on Baker Day, July 24, 1946, at Bikini Island, the second explosion in Operation Crossroads.
Source: *Boston Herald American.*

PLATE H Damage to a brick house by an overpressure of 5 pounds per square inch, the amount of explosive blast found about 4 miles from ground zero of a 1-megaton nuclear explosion. This house was destroyed by an atomic bomb at the Nevada Test Site.

Source: Nuclear Defense Agency.

The shock wave produced by a nuclear explosion is called *blast*. It results from the explosion driving air outward from its center, producing changes in air pressure, known as *overpressure*, and strong winds. The overpressure can crush objects, the winds toss them about. Most damage to buildings is caused by overpressure. People and objects like trees and telephone poles are destroyed by the winds.

Overpressure is measured in pounds per square inch (psi). An overpressure of a few psi can produce a tremendous amount of damage. Four miles away from a 1-megaton explosion, the overpressure would be about 5 psi, enough to reduce an unreinforced brick house to a pile of rubble (see plate H). Winds at that distance from *ground zero*—the spot directly beneath the point of detonation—would exceed 160 miles per hour.[20] While 5 psi is not enough to crush a person, people would be killed by collapsing buildings. The winds would hurl people through the air at high speeds, and they also would be struck by flying objects.

Nuclear war planners classify targets in terms of their *hardness*, that is, their ability to withstand overpressure. An underground missile silo is a *hard target*, while cities and most industrial facilities are *soft*. To *crack*—or *kill*—a hard target, nuclear tacticians would use a *ground burst:* they would detonate the bomb at ground level near or on top of the target. This would produce an extremely large overpressure near the target, but the overpressure would decrease rapidly as one moved away from ground zero. When attacking softer targets, as in *city-busting, air bursts* would be used because they spread the explosive blast over a larger area. Smaller overpressures are produced by air bursts, but because a larger area would be affected, the damage would be considerably greater.

Fallout consists of highly radioactive fission products. If a weapon were detonated at ground level, these fission products would collect on dust and water droplets, forming a lethal fallout cloud that would stay close to the ground and that could be carried long distances (see figure 16). An air burst would not produce a great *local fallout* hazard, since most fission products would be carried into the upper atmosphere, to descend later, much dispersed, as *global fallout*.

About 35% of the energy from a nuclear explosion is released in an intense burst of heat, or *thermal radiation*.[21] As a rule of thumb, a 1-megaton bomb would produce third-degree burns—burns which involve charring of the flesh and which require skin grafting—on unprotected people at distances of up to 5 miles. In addition to the direct effects on people, thermal radiation could ignite combustible materials. In urban areas, fires could also result from breaks in gas lines and fuel storage tanks. Under certain conditions, individual fires could combine into a *mass fire*, capable of consuming an entire area. There are two forms of mass fires: *firestorms* and *conflagrations*. In a firestorm, violent inrushing winds would feed the flames, producing extremely high temperatures. In a conflagration, the fire would spread along a front, sweeping over a large area. A report on

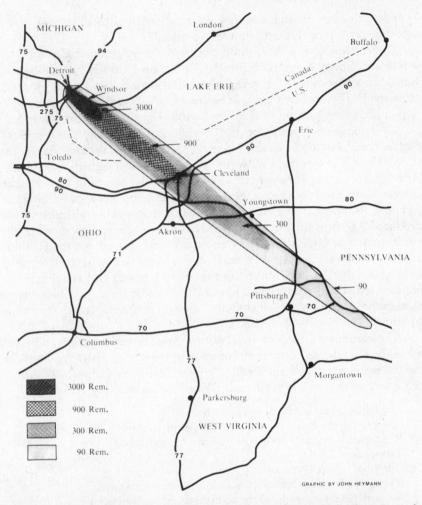

FIGURE 16 Map of fallout plume from a 1-megaton surface burst on Detroit. In the example shown, the fallout is carried by a 15-mph northwest wind. The contours indicate the regions in which people would accumulate doses of 3,000, 900, 300, and 90 rems over the 7-day period following the attack. (A dose of from 250 to 450 rems will kill 50% of the people exposed from acute radiation sickness.) Wind direction and speed have an important effect on the number of deaths from fallout. A southwest wind, for example, would carry the fallout into a region of Canada much less heavily populated than the Cleveland-Pittsburgh area.

Source: Adapted from U.S., Congress, Office of Technology Assessment, *The Effects of Nuclear War*, OTA-NS-89, May 1979, p. 25.

---

the effects of nuclear war issued by the federal Office of Technology Assessment (OTA) in 1979 said: "A firestorm is likely to kill a high proportion of the people in the area of the fire, through heat and through asphyxiation of those in shelters. A conflagration spreads slowly enough so that people in its path can escape, though a conflagration caused by a nuclear attack might take a heavy toll of those too injured to walk."[22]

Direct radiation, a short pulse of extremely dangerous radiation, also would accompany a nuclear explosion. Radiation deaths at Hiroshima and Nagasaki, where the bombs were detonated in the air, were caused by direct radiation.[23] The *neutron bomb*, or *enhanced radiation weapon*, would produce a higher proportion of direct radiation than most nuclear weapons.

Electromagnetic pulse, or *EMP*, while not physically dangerous to people, could destroy or disable electrical and electronic systems. EMP consists of short, but extremely powerful, pulses of electromagnetic waves similar to radio waves. These waves could induce brief surges of high-voltage current in electrical equipment, upsetting its operation and even burning out components. Nuclear weapons detonated at altitudes above 19 miles would produce EMP extending over thousands of miles.[24] Several weapons might be detonated at high altitudes by an attacker to destroy communications and electric power systems. A single nuclear detonation several hundred miles above the U.S. could shut down the electric power grid and knock out unprotected communications equipment from coast to coast.[25] Most communications equipment—including much military communications equipment—is not *EMP-hardened*.[26]

The combined effects of blast, fallout, heat, direct radiation and EMP would be devastating, even from the most optimistic perspective. Consider the following pep talk from a 1961 "how-to" book by Colonel Mel Mawrence and John Clark Kimball called *You Can Survive The Bomb:*

The English endured the bombing of Britain.
The French and Belgians endured the juggernaut of two world wars.
The Japanese endured the disastrous firebombing of their major cities, plus two nuclear attacks.
The Jews endured genocide.
The early Christians endured the Colosseum and catacombs.
Our own ancestors endured the wilderness and Valley Forge.
Yes, and the Russians endured, too—nearly 20 million dead in World War II. . . .
If you have the courage, here is what you may have to face:
—The fireball
—Heat
—Air blast
—The crater
—Initial and residual fallout radiation
—The aftermath, with its *nagging low-level radiation*, creating new problems to be solved, new hostile surroundings to be *conquered* only by *courage and wit*.[27]

Mawrence and Kimball's book is only one of a number of nuclear war survival manuals. Another is Richard Gerstell's 1950 Bantam paperback,

How to Survive an Atomic Bomb. Gerstell was a radiological safety specialist at Operation Crossroads. He presented his readers with a number of simple rules to follow in case of nuclear attack. The most important piece of advice in the book was the following maxim: *"Everyone must always lie down full-length on his stomach with his face buried in his arms"* (original emphasis).[28] If you were walking along a city street and saw the flash, you should assume this position next to the nearest building. Farmers were told to dive into the nearest plowed furrow if nothing better was at hand.[29]

Gerstell downplayed the hazards of radiation and fallout, saying that people should not let words scare them:

> The reason why so many of us have been thinking that the radiation is worse than it is, is because there's been so much *loose talk* about the atomic bomb and the rays it makes. We've been scared by those two words— "radiation" and "radioactivity." It's time we all stopped jumping when we hear them and learned a few plain *facts* about them. . . .[30]

One of Gerstell's *facts* was that "because they have worked and experimented with radioactivity for over fifty years, doctors know more about sickness from atomic radiation than they know about polio, colds, and some other common diseases."[31] This argument confused the sheer quantity of information gathered on radiation with knowledge or understanding that would be useful in ameliorating its effects. Taking Gerstell's example, doctors, despite their allegedly inferior knowledge about polio, developed a successful vaccine for polio just four years after the publication of Gerstell's book. There is no "vaccine" against radiation sickness.

Gerstell recommended that everyone take at least one shower a day after an attack as a precautionary measure, being sure to use plenty of soap and water.[32] If *"hot mist"* or *"hot dust"* should appear after an explosion, Gerstell told his readers to "clean your house working from top to bottom, as any good housewife would do."[33] While carrying out this tedious cleaning, Gerstell suggested that one always remember that his rules "will help to keep you *safe.*"[34]

During the early 1950s, fears about nuclear war created a market for atomic-age real estate. In July 1950, Washington, D.C. realty ads featured homes "a *safe* 58 miles from Washington" and "small farms out beyond the range of A-bombs."[35] America's inventors also responded to the demand for protection. One 1951 inventor displayed a quilted suit of khaki pajamas made for his six-year-old son. It contained 5 pounds of shredded lead. (The average weight of a 6-year-old is about 47 pounds.) The AEC said that the suit was the first with real possibilities for radiation protection.[36]

In response to the Soviet Union's successful test of an atomic bomb in 1949, the U.S. government established a *civil defense (CD)* program designed to reduce the damage from a nuclear attack. Originally, plans called for the speedy *evacuation* of cities as soon as radar picked up approaching

bombers, and suggested that people *"duck and cover."*[37] The invention of
H-bombs, with their far more powerful yields, made the duck-and-cover
program obsolete. H-bombs also raised the possibility that huge areas of the
country would be covered by fallout. As the Office of Technology Assessment pointed out in 1979, advances in technology rendered the first CD
plans useless.

> Previously, civil defense could be conceptualized as moving people a short
> distance out of cities, while the rest of the country would be unscathed and
> able to help the target cities. Fallout meant that large areas of the
> country—the location of which was unpredictable—would become con-
> taminated, people would be forced to take shelter in those areas, and their
> inhabitants, thus pinned down, would be unable to offer much help to
> attacked cities for several weeks.
>
> The advent of ICBMs necessitated further changes. Their drastically
> reduced *warning times* precluded evacuations on radar warning of at-
> tack.[38]

After the Berlin crisis in 1961, the Kennedy Administration pushed a
major program to set up *fallout shelters*, which were designed to protect
people from radiation, though they offered little protection against blast.[39]
This effort was the last big drive toward a large-scale CD program.

In its 1979 report, the Office of Technology Assessment estimated the
effects of several hypothetical attacks designed to destroy cities and industrial capacity in the U.S. and the Soviet Union. The OTA concluded that
using only ten MIRVed missiles, either superpower could shatter the other's
economy. This could most readily be accomplished by destroying oil re-
fineries, vital industrial facilities that are expensive, technologically complex, and would take years to rebuild. Using only ten MIRVed missiles, the
USSR could knock out 64% of U.S. oil-refining capacity.[40] A U.S. attack
could destroy 73% of Soviet refining capacity.[41] Both sides are capable of
mounting such attacks many times over.

The OTA also looked at the possible effects of city-busting strikes
against Detroit and Leningrad, both large industrial cities with populations
of about 4.3 million. The agency estimated the impact of a number of *attack
scenarios:* a 1-megaton ground burst on Detroit; a 1-megaton air burst on
Detroit; a 25-megaton air burst on Detroit; a 1-megaton air burst on Leningrad; a 9-megaton air burst on Leningrad; and an attack on Leningrad using
ten 40-kiloton weapons. The number of deaths and injuries that would
result from each attack is shown in figure 17.

A 1-megaton ground burst on downtown Detroit would carve a crater
200 feet deep and 1,000 feet wide out of the center of the city. A rim of
highly radioactive dirt about 2,000 feet wide would surround the crater. No
structures would remain standing within 1.7 miles of ground zero. Virtually

FIGURE 17 SUMMARY OF EFFECTS OF HYPOTHETICAL NUCLEAR ATTACKS
ON DETROIT AND LENINGRAD

| | DETROIT | |
	Dead	Injured
1-megaton ground burst	220,000	420,000
1-megaton air burst	470,000	630,000
25-megaton air burst	1,840,000	1,360,000
	LENINGRAD	
1-megaton air burst	890,000	1,260,000
9-megaton air burst	2,460,000	1,100,000
Ten 40-kiloton air bursts	1,020,000	1,000,000

Source: U.S., Congress, Office of Technology Assessment, *The Effects of Nuclear War*, OTA-NS-89, May 1979, p. 37.

everyone that close to ground zero would be killed.[42]

In the region 1.7 to 2.7 miles from ground zero, typical multistory residential and commercial buildings would have their walls blown completely out. Proceeding outward, more and more buildings would have their skeletons remaining. Many people would be killed by collapsing buildings. Few within 2 miles of ground zero would survive the blast.[43]

In the ring 2.7 to 4.7 miles from ground zero, buildings with thin walls would have the contents of their upper floors blown out onto the streets. Windows, frames, and interior partitions would be blown out of large buildings. Low residential buildings would be totally destroyed or heavily damaged. Fires would start and would continue to spread for at least 24 hours. About 5% of the population in this area would be killed, and nearly half would be injured.[44]

From 4.7 to 7.4 miles out, few people would be killed, but about 25% would be injured. Buildings would suffer light to moderate damage. Enough burn injuries would occur in this region alone to overload the entire country's burn-treatment centers. Depending on weather conditions, between 3,000 and 75,000 people would sustain burns requiring special medical attention (see plate I). As of 1977, there were only eighty-five specialized burn centers in the U.S., with facilities for a total of 1,000 to 2,000 patients.[45]

The OTA's estimate of 640,000 casualties—combined deaths and injuries—does not include the toll from fallout. The direction of the wind would determine how many people would die from radiation poisoning (see figure 16).

Without immediate attention, a large number of additional deaths would occur as a result of inadequate medical care. Hospitals in the region

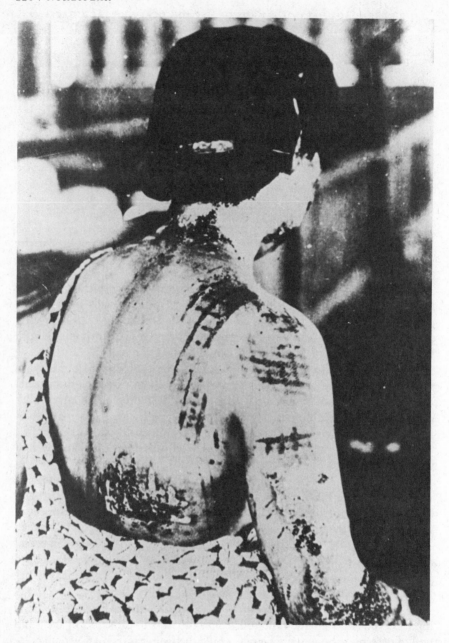

PLATE I Burn victim from the Hiroshima bombing. The woman's skin is burned in a pattern corresponding to the dark portions of a kimono she wore at the time of the attack. Flash burns from thermal radiation (heat) can affect people miles away from ground zero.

Source: Department of Defense.

would be unable to help, since 55% of the area's hospital capacity would have been destroyed and another 15% severely damaged.[46] Rescue operations would have to be organized from outside. They would be hampered by the threat of fallout and the destruction of much of Detroit's transportation system. Once the injured began to be moved out of the area, the entire country's medical facilities would be overburdened. In 1977 there were only 1.4 million hospital beds in the nation.[47]

Figure 18 shows the overpressure rings that would result from a 1-megaton and a 25-megaton air burst on Detroit. The death tolls from these attacks would be higher than for the ground burst, although close to ground zero the destruction would be less complete. No crater would be formed, and there would be no *close-in fallout*.[48]

The hypothetical attack on Leningrad would have equally disastrous effects. Because Leningrad is more densely populated than Detroit, about twice the number of casualties would result from the 1-megaton air burst, and a 9-megaton air burst would kill more people than a 25-megaton air burst on Detroit. Most of the housing in Leningrad consists of high-rise apartment buildings, and at about 5 psi overpressure, their walls would be destroyed, sweeping the occupants into the streets.[49] Figure 19 shows a targeting plan and the resulting overpressures for an attack on Leningrad with ten 40-kiloton weapons.

As bleak as these hypothetical attacks on single cities sound, they are minor disasters compared with a more widespread nuclear war. The initial survivors of a general attack would have no one to rescue them, since everyone would be busy trying to rescue himself. Fallout would force people into shelters for up to a month. There would be severe shortages of medicine and doctors. Most of the injured would go without medical treatment. Habitable housing would be hard to find in some regions, adding to the death toll, especially if the attack occurred during the winter. Food would be in short supply, as would fuel. Whole industries would be obliterated. The social fabric would be torn to shreds.

The OTA estimated the consequences of major nuclear attacks on both the U.S. and the USSR designed to destroy industrial and military capacity. The attacks would kill up to 165 million Americans[50] and as many as 100 million Soviet citizens.[51] These estimates include only deaths during the first thirty days after the attacks; they do not take into account deaths that would occur from economic disruption and among the injured. Conditions would get worse before they got better. Achieving economic recovery after the attack would be very difficult. The OTA concluded that it is likely that the survivors would return to "the economic equivalent of the Middle Ages."[52]

The OTA pointed out that a nuclear war would have long-term effects as a result of low-level exposure to ionizing radiation and other disruptions to the environment. Some scientists believe that a large number of nuclear

Detroit—1 Megaton and 25 Megatons Air Burst

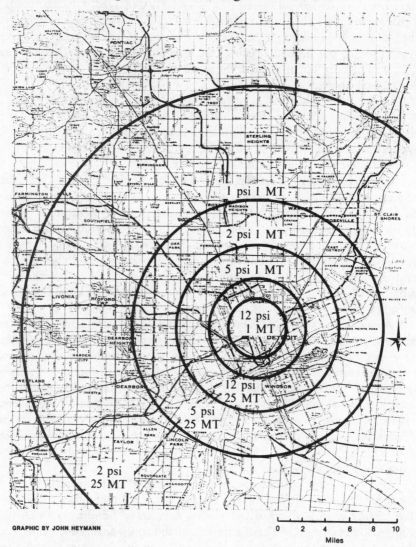

1 psi 1 MT

2 psi 1 MT

5 psi 1 MT

12 psi
1 MT

12 psi
25 MT

5 psi
25 MT

2 psi
25 MT

GRAPHIC BY JOHN HEYMANN

```
0    2    4    6    8    10
              Miles
```

FIGURE 18 Comparison of the overpressures resulting from 1-megaton and 25-megaton air bursts over Detroit. The rings indicate the strength of the blast with which the enclosed area would be hit. For both the 1-megaton and the 25-megaton cases, rings are drawn at the 12, 5, and 2 psi limits. The 1 psi ring is shown for the 1-megaton attack. In the case of the 25-megaton attack, the 1 psi ring lies completely off the map, 30.4 miles from ground zero.

Source: Adapted from U.S., Congress, Office of Technology Assessment, *The Effects of Nuclear War*, OTA-NS-89, May 1979, p. 44.

Leningrad — Ten 40-Kiloton Air Burst

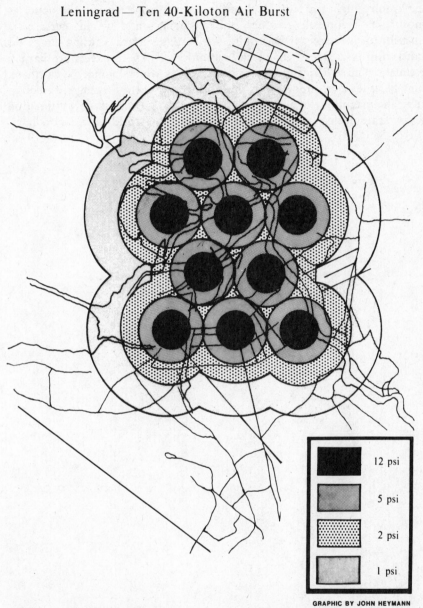

■	12 psi
▨	5 psi
▦	2 psi
▢	1 psi

GRAPHIC BY JOHN HEYMANN

FIGURE 19 Hypothetical attack on the city of Leningrad with ten 40-kiloton weapons. The bomb pattern has been selected such that the 5 psi overpressure rings are just touching. The Office of Technology Assessment estimates that such an attack would kill 1,020,000 people and would injure another million.

Source: Adapted from U.S., Congress, Office of Technology Assessment, *The Effects of Nuclear War,* OTA-NS-89, May 1979, p. 44.

explosions might seriously deplete the earth's ozone layer in the upper atmosphere. This could produce climate changes and would increase the amount of ultraviolet light striking the earth's surface, causing dangerous burns and widespread ecological disruptions.[53] These effects are hard to estimate. The OTA cautioned that the effects which cannot be calculated may be at least as important as those which the agency attempted to quantify. The magnitude of a nuclear holocaust is so great as to defy estimation. A full-scale nuclear war would be the end of the world as we know it.

There Is a Choice

No one wants a nuclear war, and no one would win should one occur. Throughout history, however, wars have been fought which no one wanted. Nuclear weapons do not build themselves. They are the product of a tremendous amount of human ingenuity, money, and labor. They exist only because people in the nuclear nations believe they need to build them.

Part VII demonstrated the way the nuclear mindset has limited the realm of possibility in energy planning. Widely held beliefs about energy growth, conservation, and solar power led people to conclude that nuclear energy was necessary and cheap. When this mindset was translated into policy, choice was artificially constrained. The alternatives to a nuclear-powered future were shortchanged.

The nuclear arms race is also mindset translated into policy. Human choices drive the arms race. Probably the greatest obstacle to preventing nuclear war is the belief that nuclear weapons are necessary, that there is no choice but to continue to build and stockpile them. Recognizing that the nuclear arms race is a product of our mindset is the first step toward widening the realm of choice for nuclear arms control.

Notes

Footnotes are listed by chapter, and the epigrams for each part are listed at the beginning of the first chapter of that part.

PART I: NUCLEAR VISIONS

1. Albert Einstein, fund-raising telegram for the Emergency Committee of Atomic Scientists, 23 May 1946, in *Einstein On Peace,* Otto Nathan and Heinz Norden, eds. (New York: Simon and Schuster, 1960), p. 376.

CHAPTER 1: THE ELIXIR OF LIFE

2. *New York Times,* 16 February 1896, p. 4.
3. Percy Brown, *American Martyrs to Science Through the Roentgen Rays* (Springfield, Illinois: Charles C. Thomas, 1936), p. 38.
4. *New York Times,* 29 May 1898, p. 14.
5. Otto Glasser, *William Conrad and the Early History of the Roentgen Rays* (Springfield, Illinois: Charles C. Thomas, 1934), p. 44.
6. Ibid.
7. *New York Times,* 24 July 1911, p. 1.
8. *New York Times,* 8 May 1931, p. 27.
9. Ibid.
10. *New York Times,* 25 November 1904, p. 6.
11. Frederick Soddy, *The Interpretation of Radium and the Structure of the Atom* (London: John Murray, 1920), p. 27 (first published 1909).
12. Lawrence Badash, *Radioactivity in America—Growth and Decay of a Science* (Baltimore: The Johns Hopkins University Press, 1979), p. 26.
13. *New York Times,* 5 July 1922, p. 16.
14. "Dr. Morton's Theory of the Therapeutic Value of Radium Solutions," *Scientific American,* vol. XC, no. 5, 30 January 1904, p. 90.
15. *New York Times,* 6 February 1904, p. 6.
16. *Technology Review,* vol. 6, no. 2, April 1904, p. 249.
17. *New York Times,* 6 February 1904, p. 6.
18. *Technology Review,* vol. 6, no. 2, April 1904, p. 249.

19. *New York Times,* 28 May 1905, III, p. 6.
20. *New York Times,* 21 June 1905, p. 1.
21. *New York Times,* 25 October 1914, p. 8.
22. Howard A. Kelly, "Radium in Surgery," *Journal of the American Medical Association,* vol. LX, no. 20, 17 May 1913, p. 1574.
23. Douglas C. Moriarta, "Radium: A Recognition of Its Efficiency and a Plea for a More Thorough Investigation," *Radium,* vol. VII, no. 2, May 1916, p. 59.
24. Harold Swanberg, *Radiologic Maxims* (Quincy, Illinois: Radiologic Review Publishing Co., 1932), p. 113.
25. Ibid., p. 77.
26. Ibid., p. 78.
27. Ibid., p. 72.
28. Ibid., p. 116.
29. Ibid., p. 78.
30. *New York Times,* 28 December 1908, p. 7.
31. Brown, *American Martyrs,* p. 28.
32. *Scientific American Supplement,* 24 July 1909, p. 54.
33. *New York Times,* 31 July 1897, p. 6.
34. William Rollins, "The Effects of X-Light on The Crystalline Lens," *Boston Medical and Surgical Journal,* vol. 148, no. 14, p. 364.
35. Brown, *American Martyrs,* p. 92.
36. *New York Times,* 20 January 1925, p. 1.
37. Ibid.
38. Ibid.
39. Ibid.
40. *New York Times,* 28 April 1928, p. 21.
41. *New York Times,* 20 January 1925, p. 1.
42. Ibid.
43. *New York Times,* 23 May 1928, p. 11.
44. *New York Times,* 5 June 1928, p. 1.
45. *New York Times,* 14 June 1928, p. 9.
46. *New York Times,* 5 June 1928, p. 1.

CHAPTER 2: THE WORLD SET FREE

1. Frederick Soddy, *The Interpretation of Radium and the Structure of the Atom* (London: John Murray, 1920), p. 172 (first published 1909).
2. Ibid., p. 173.
3. H. G. Wells, *The World Set Free* (London: MacMillan and Co., Limited, 1914), dedication page.
4. Ibid., p. 203.
5. Ibid., pp. 103–104.
6. Ibid., p. 38.
7. Ibid., p. 36.
8. Ibid., p. 39.

9. Ibid., p. 101.

10. Ibid., pp. 221–223.

11. Spencer R. Weart and Gertrude Weiss Szilard, eds., *Leo Szilard: His Version of the Facts* (Cambridge, Massachusetts, and London, England: The MIT Press, 1978), pp. 12, 16.

12. Ibid., p. 17.

13. Bernard T. Feld and Gertrude Weiss Szilard, eds., *The Collected Works of Leo Szilard: Scientific Papers* (Cambridge, Massachusetts, and London, England: The MIT Press, 1972), p. 529.

14. Weart and Szilard, *His Version*, p. 17.

15. Feld and Szilard, *Collected Works*, pp. 607–621.

16. Weart and Szilard, *His Version*, p. 18.

17. Ibid.

18. Feld and Szilard, *Collected Works*, p. 734.

19. Ibid.

20. Weart and Szilard, *His Version*, p. 41.

21. Ibid., p. 42.

22. Ibid., p. 53.

23. Ibid., p. 57.

24. Ibid.

25. Louis A. Turner, "Nuclear Fission," *Review of Modern Physics*, vol. 12, no. 1, January 1940, p. 1.

26. Feld and Szilard, *Collected Works*, p. 222.

27. Weart and Szilard, *His Version*, pp. 115–116.

28. Feld and Szilard, *Collected Works*, p. 173.

29. William L. Laurence, *Men and Atoms* (New York: Simon and Schuster, 1959), p. 7.

30. Ibid., p. 9.

31. *New York Times*, 25 February 1939, p. 17.

32. *New York Times*, 30 April 1939, p. 35.

33. Ibid.

34. R. M. Langer, "Fast New World," *Colliers*, vol. 106, no. 1, 6 July 1940, p. 19.

35. Ibid., p. 54.

36. Ibid., p. 18.

37. Ibid.

38. Ibid.

39. Ibid., p. 19.

40. Ibid., p. 54.

41. Ibid.

42. Ibid.

43. John J. O'Neill, "Enter Atomic Power," *Harper's*, vol. 181, June 1940, p. 3.

44. Ibid., p. 7.

45. William L. Laurence, "The Atom Gives Up," *Saturday Evening Post*, vol. 213, no. 10, 7 September 1940, p. 12.

46. Ibid., p. 62.

47. Ibid., p. 12.

48. Ibid.

49. Ibid., p. 60.

50. Ibid.

51. Ibid., p. 63.

52. Ibid., p. 62.

53. Ibid.

CHAPTER 3: THE BATTLE OF THE LABORATORIES

1. Richard G. Hewlett and Oscar E. Anderson, *The New World, 1939/46: A History of the United States Atomic Energy Commission*, 2 vols. (University Park, Pennsylvania: The Pennsylvania State University Press, 1962), vol. 1, p. 2.

2. Lawrence Badash, Joseph O. Hirschfelder, and Herbert P. Broida, *Reminiscences of Los Alamos* (Dordrecht, Holland, Boston, U.S.A., and London, England: D. Reidel Publishing Company, 1980), p. 67.

3. Spencer R. Weart and Gertrude Weiss Szilard, eds., *Leo Szilard: His Version of the Facts* (Cambridge, Massachusetts, and London, England: The MIT Press, 1978), p. 84.

4. Ibid., pp. 94–95.

5. Ibid., 95.

6. Ibid.

7. Hewlett and Anderson, *The New World*, p. 17.

8. Weart and Szilard, *His Version*, pp. 95–96.

9. Leslie R. Groves, *Now It Can be Told* (New York and Evanston: Harper & Row, 1962), p. 48.

10. Ibid., p. 253.

11. Hewlett and Anderson, *The New World*, pp. 205–206.

12. William L. Laurence, *Dawn Over Zero* (New York: Alfred A. Knopf, 1946), p. 127.

13. Ibid.

14. Groves, *Now It Can Be Told*, p. 70.

15. William L. Laurence, *Men and Atoms* (New York: Simon and Schuster, 1959), p. 131.

16. U.S., Department of State, *The International Control of Atomic Energy: Growth of a Policy*, publication 2702, "Statement by the Secretary of War, August 6, 1945," appendix no. 2, p. 98.

17. Groves, *Now It Can Be Told*, p. 140.

18. Ibid.

19. Ibid.

20. Ibid.

21. Laurence, *Men and Atoms*, p. 52.

22. Hewlett and Anderson, *The New World*, p. 238.

23. Martin J. Sherwin, *A World Destroyed* (New York: Vintage Books, 1977), p. 63 (first published by Alfred A. Knopf, Inc., in 1975).

24. Peter Goodchild, *J. Robert Oppenheimer: Shatterer of Worlds* (Boston: Houghton Mifflin Company, 1981), pp. 87–92.

25. Daniel Lang, *From Hiroshima to the Moon: Chronicles of Life in the Atomic Age* (New York: Simon and Schuster, 1959), p. 1.
26. Ibid., p. 31.
27. Ibid., p. 33.
28. Arthur Holly Compton, *Atomic Quest: A Personal Narrative* (New York: Oxford University Press, 1956), p. 10.
29. Groves, *Now It Can Be Told*, p. 100.
30. Ibid., p. 146.
31. Ibid.
32. Ibid.
33. Ibid., p. 191.
34. Ibid., p. 13.
35. Ibid., p. 17.
36. Ibid., p. 325.
37. Ibid.
38. William L. Laurence, *Men and Atoms*, p. 48.
39. Groves, *Now It Can Be Told*, p. 191.
40. Compton, *Atomic Quest*, p. 182.
41. Lang, *From Hiroshima to the Moon*, p. 9.
42. Ibid., p. 6.
43. Albert Wohlstetter et al., *Swords From Plowshares* (Chicago: The University of Chicago Press, 1979), p. 69.
44. Hewlett and Anderson, *The New World*, p. 761.
45. "When Uranium Was 'Tuballoy'", *Nucleonics,* vol. 12, no. 10, October 1954, p. 21.
46. Hewlett and Anderson, *The New World*, p. 228.
47. Goodchild, *Shatterer of Worlds*, p. 77.
48. Groves, *Now It Can Be Told*, p. 147.
49. Hewlett and Anderson, *The New World*, p. 231.
50. Badash, Hirschfelder, and Broida, *Reminiscences of Los Alamos*, p. 140.
51. Ibid.
52. Eleanor Jette, *Inside Box 1663* (Los Alamos, N.M.: Los Alamos Historical Society, 1967), cited in Goodchild, *Shatterer of Worlds*, p. 79.
53. Goodchild, *Shatterer of Worlds*, p. 6.
54. Ibid.
55. Ibid.
56. U.S., House Committee on Interstate and Foreign Commerce, Subcommittee on Health and the Environment, *Effects of Radiation on Human Health—Health Effects of Ionizing Radiation,* Serial No. 95-179, Ninety-Fifth Congress, 2nd Session, January 24, 25, 26, February 8, 9, 10, and 28, 1978, Volume 1, attachment "Announced United States Nuclear Detonations, January 1, 1978," testimony of Donald M. Kerr, Acting Assistant Secretary for Defense Programs, Department of Energy, p. 341.
57. Hewlett and Anderson, *New World*, p. 386.
58. Goodchild, *Shatterer of Worlds*, p. 145.
59. Edward Teller with Allen Brown, *The Legacy of Hiroshima* (Garden City, New York: Doubleday & Co., 1962), p. 17.

60. Henry DeWolf Smyth, *Atomic Energy for Military Purposes, The Official Report on the Development of the Atomic Bomb under the Auspices of the United States Government, 1940–1945* (Princeton: Princeton University Press, 1945), pp. 253–254.
61. Laurence, *Men and Atoms,* p. 118.
62. Kenneth T. Bainbridge, "A Foul and Awesome Display," *The Bulletin of the Atomic Scientists,* vol. XXXI, no. 5, May 1975, p. 46.
63. Ibid.
64. Groves, *Now It Can Be Told,* p. 301.
65. Laurence, *Men and Atoms,* p. 112.
66. Groves, *Now It Can Be Told,* pp. 299–301.
67. Compton, *Atomic Quest,* p. 214.
68. Goodchild, *Shatterer of Worlds,* p. 151.
69. Hewlett and Anderson, *The New World,* p. 352.
70. Ibid., p. 383.
71. Ibid., p. 386.
72. Ibid.
73. Ibid., pp. 372–373.
74. U.S., Department of State, *The International Control of Atomic Energy: Growth of a Policy,* publication 2701, appendix 1, "Statement by the President of the United States, August 6, 1945," p. 95.
75. Groves, *Now It Can Be Told,* p. 331.
76. Ibid.
77. U.S., Department of State, *The International Control of Atomic Energy: Growth of a Policy,* appendix 1, p. 95.
78. Ibid., p. 96.
79. Ibid.
80. Ibid., p. 95.
81. Groves, *Now It Can Be Told,* p. 328.

CHAPTER 4: ATOMS FOR PEACE

1. David E. Lilienthal, *Change, Hope, and the Bomb* (Princeton, New Jersey: Princeton University Press, 1963), pp. 109–110.
2. U.S., Department of State, *The International Control of Atomic Energy: Growth of a Policy,* publication 2702, appendix 2, "Statement by the Secretary of War, August 6, 1945," p. 104.
3. William L. Laurence, "Paradise or Doomsday?", *Woman's Home Companion,* May 1948, p. 33.
4. Ibid.
5. Ibid., p. 75.
6. Ibid., p. 74.
7. Heinz Haber, *Our Friend the Atom* (New York: Simon and Schuster, 1956), p. 13.
8. Ibid., pp. 20–21.
9. Ibid., p. 160.

10. U.S., Department of State, *A Report on the International Control of Energy* (Garden City, New York: Doubleday & Company, 1946), p. 37.

11. Ibid., p. 39.

12. Ibid., p. 21.

13. Ibid., p. 34.

14. Ibid., p. 21.

15. Richard G. Hewlett and Francis Duncan, *Atomic Shield, 1947/1952, vol. II: A History of the United States Atomic Energy Commission* (University Park, Pennsylvania, and London: The Pennsylvania University Press, 1969), pp. 676–677.

16. *New York Times,* 9 December 1953, p. 2.

17. Ibid.

18. U.S., Atomic Energy Commission press release, remarks prepared for delivery at Founders' Day Dinner, National Association of Science Writers, 16 September 1954, p. 9.

19. "A Wand Wave, A New Era," *Life,* vol. 37, no. 12, 20 September 1954, p. 141.

20. "Commercial Electric Power From Atomic Energy," *Science,* vol. 122, no. 3161, 29 July 1955, p. 192.

21. "Opening Remarks by Dag Hammarskjöld," *Proceedings of the International Conference on the Peaceful Uses of Atomic Energy,* 8–20 August 1955, Geneva, record of the conference published by the United Nations (New York, 1956), vol. 16, sess. 1, p. 28.

22. Atomic Industrial Forum, *Forum Memo,* vol. 6, no. 8, August 1959, p. 22.

23. Atomic Industrial Forum, *Forum Memo,* vol. 8, no. 5, May 1961, p. 11.

24. *Nucleonics,* vol. 19, no. 5, May 1961, p. 21.

25. Atomic Industrial Forum, *Forum Memo,* vol. 5, no. 8, August 1958, p. 36.

26. *Nucleonics,* "The Nuclear Space Age," vol. 19, no. 4, April 1961, p. 156.

27. Ibid.

28. Robert E.L. Adamson, "Washington Foresees Major Role for Nuclear Space Power," *Nucleonics,* vol. 19, no. 4, April 1961, p. 55.

29. Wright H. Langham, "Radiation Safety in the Development and Use of Nuclear Energy for Rocket Propulsion," *Health Physics,* vol. 8, 1962, p. 305.

30. U.S., Atomic Energy Commission, *Annual Report to Congress of the AEC for 1970,* January 1971, p. 319.

31. "Statement by Frank DiLuzio," *Proceedings of the Atomic Industrial Forum's 1964 Annual Conference,* 30 November–3 December 1964, San Francisco (New York, 1965), vol. 3, p. 195.

32. U.S., Atomic Energy Commission, *1972 Annual Report to Congress, Operating and Development Function,* 31 January 1973, p. 25.

33. John McPhee, *The Curve of Binding Energy* (New York: Farrar, Straus and Giroux, 1973), pp. 172–174.

34. Ibid., p. 175.

35. Freeman Dyson, *Disturbing the Universe* (New York: Harper & Row, Publishers, 1979), p. 112.

36. Ibid., p. 114.

37. Ibid., p. 115.

38. Ibid., p. 113.

39. McPhee, *The Curve of Binding Energy,* p. 183.

40. Ibid., p. 184.

41. Glenn T. Seaborg and William R. Corliss, *Man and Atom* (New York: E. P. Dutton and Co., Inc., 1971), p. 174.

42. Ibid.

43. Edward Teller, "Peaceful Uses of Fusion," *Proceedings of the Second United Nations International Conference on the Peaceful Uses of Atomic Energy,* 1–13 September 1958, vol. 31: *Theoretical and Experimental Aspects of Controlled Nuclear Fusion* (Geneva: United Nations, 1958), P/2410 USA, p. 32.

44. Seaborg and Corliss, *Man and Atom,* p. 188.

45. Ibid., pp. 188–194.

46. U.S., Congress, Joint Committee on Atomic Energy, *Commercial Plowshare Services and Related Background Material (H.R. 18448 and S.3783),* hearings before the Subcommittee on Legislation, 90th Congress, 2d sess., 19 July 1968, appendix 2, "Remarks by Congressman Hosmer to the Commonwealth Club of California, July 5, 1968," pp. 38–43.

47. Ibid., p. 42.

48. Ibid.

49. Seaborg and Corliss, *Man and Atom,* p. 184.

50. U.S., Congress, Joint Committee on Atomic Energy, *AEC Authorizing Legislation, Fiscal Year 1974,* hearings before the JCAE, 93rd Congress, 1st sess., 30 January 1973, part I, p. 76.

51. Albert Wohlstetter et al., *Swords From Plowshares* (Chicago: University of Chicago Press, 1979), forward by Fred Charles Iklé, p. viii.

PART II: THE MINISTRY OF TRUTH

1. Mike Levin (National Security Agency), "Problems of Intelligence and Security in a Democratic Society," *Classification Management,* vol. XV, 1979, p. 124.

2. Frank B. Shants, "Countering the Anti-Nuclear Activists," *Public Relations Journal,* vol. 34, no. 10, October 1978, p. 10.

CHAPTER 5: THE SECRECY SYSTEM

3. U.S., The President, Executive Order 12065: "National Security Information," section 1-2, 28 June 1978.

4. Interview of Murray L. Nash, deputy director of the Department of Energy's Office of Classification, Germantown, Maryland, 24 March 1980.

5. Alvin M. Weinberg, "Social Institutions and Nuclear Energy," *Science,* vol. 177, no. 4043, 7 July 1972, p. 34.

6. J. Gustave Speth, Arthur R. Tamplin and Thomas B. Cochran, "Plutonium Recycle: The Fateful Step," *Bulletin of the Atomic Scientists,* vol. XXX, no. 9, November 1974, p. 20.

7. U.S., Department of Energy, *Classification of Information Manual,* 5650.2, 12 December 1972, p. I-5.

8. Nash interview.

9. DOE, *Classification of Information Manual,* p. VII-1.

10. Ibid., p. VII-2.
11. "News and Notes," *Science*, vol. 121, no. 3152, 27 May 1955, p. 756.
12. U.S., Interagency Classification Review Committee, Information Security Oversight Office, "National Security Information," *Federal Register*, vol. 43, no. 194, part V, 5 October 1978, p. 46283.
13. Nash interview.
14. DOE, *Classification of Information Manual*, p. IV-7.
15. Interview with government employee who requested not to be identified, conducted in 1980.
16. U.S., The President, Executive Order 12065, section 6-104.
17. Ibid.
18. *Atomic Energy Act of 1946.*
19. U.S., Department of Energy, Assistant Secretary for Defense Programs, Office of Classification, *Guidebook for the Authorized Classifier*, DOE/DP-0008, p. 2.
20. V. L. Parsegian, "Is There an Atomic Curtain?", *New York Times Magazine*, 14 October 1956, p. 72.
21. Howard Morland, "The H-bomb secret, To know how is to ask why," *The Progressive*, vol. 43, November 1979, p. 15.
22. Ibid.
23. DOE, *Classification of Information Manual*, p. V-1.
24. Nash interview.
25. David Wise, *The Politics of Lying* (New York: Vintage Books, 1973), pp. 80–81.
26. U.S., Nuclear Regulatory Commission, Division of Security, Office of Administration, *NRC Security: Your Responsibilities*, 1 December 1978, p. 9.
27. U.S., DOE, *Guidebook for the Authorized Classifier*, DOE/DP-0008, p. 5.
28. Interview by telephone of Marlene Flor, Department of Energy, Office of Classification, 27 March 1980.
29. DOE, *Classification of Information Manual*, p. VII-2.
30. Ibid., p. IV-10.
31. Ibid., p. VI-7.
32. Ibid.
33. Ibid., p. VI-10.
34. Louis N. Ridenour, "The Hydrogen Bomb," *Scientific American*, vol. 182, no. 3, March 1950, p. 11.
35. Morland, "The H-bomb secret," p. 16.
36. Ibid.
37. *New York Times*, 26 March 1979, p.19.
38. *New York Times*, 10 March 1979, p. 1.
39. David E. Lilienthal, *Change, Hope and the Bomb* (Princeton, New Jersey: Princeton University Press, 1963), p. 29.
40. *New York Times*, 26 March 1979, p. 19.
41. Ibid.
42. Howard Morland, "Errata," *The Progressive*, vol. 43, December 1979, p. 36.
43. Michael Macdonald Mooney, " 'Right Conduct' For A 'Free Press,' " *Harper's*, vol. 260, March 1980, p.42.

44. *New York Times,* 1 April 1979, p. 9.
45. *New York Times,* 25 March 1979, p. E18.
46. *New York Times,* 29 March 1979, p. 22.
47. *New York Times,* 15 May 1979, p. 18.
48. *New York Times,* 5 September 1979, p. A18.
49. Laurence H. Tribe and David H. Remes, "Some Reflections on The Progressive Case: publish *and* perish?", *The Bulletin of the Atomic Scientists,* vol. 36, no. 3, March 1980, p. 23.
50. *New York Times,* 20 September 1979, p. 23.
51. Ibid.
52. Samuel H. Day, Jr., "The Other Nuclear Weapons Club: How the H-bomb Amateurs Did Their Thing," *The Progressive,* vol. 43, November 1979, pp. 33–34.
53. *New York Times,* 18 September 1979, p. 1.
54. Mooney, "'Right Conduct,'" p. 36.
55. Ibid.
56. "Does The Progressive Have a Case?," *Columbia Journalism Review,* May/June 1979, p. 27.
57. *New York Times,* 18 October 1979, p. A1.
58. Tribe and Remes, "Some Reflections on The Progressive Case," pp. 23–24.
59. Ibid., p. 24.

CHAPTER 6: POLLS, ADS, AND PR

1. William A. Shurcliff, *Bombs at Bikini, The Official Report of Operation Crossroads* (New York: W. H. Wise, 1947), pp. 2, 102.
2. *New York Times,* 28 June 1946, p. 15.
3. Shurcliff, *Bombs at Bikini,* p. 36.
4. Ibid., p. 37.
5. Ibid., p. 38.
6. Ibid., appendix A, "Target Vessels, Test A," pp. 190–191.
7. Ibid., p. 139.
8. Ibid., p. 25.
9. *New York Times,* 22 July 1946, p. 1.
10. Shurcliff, *Bombs at Bikini,* p. 167.
11. "What Science Learned at Bikini," *Life,* vol. 23, no. 6, 11 August 1947, p. 78.
12. *New York Times,* 8 August 1946, p. 4.
13. "What Science Learned at Bikini," p. 79.
14. Shurcliff, *Bombs at Bikini,* p. 168.
15. David Bradley, *No Place To Hide* (Boston: Little, Brown and Company, 1948), p. 104.
16. Ibid., p. 125.
17. Ibid., p. 101.
18. U.S., Atomic Energy Commission, *Annual Report to Congress of the Atomic Energy Commission for 1969,* January 1970, p. 221.

19. U.S., Atomic Energy Commission, *Twenty-fourth Semiannual Report of the Atomic Energy Commission,* July 1958, p. 120.

20. U.S., Atomic Energy Commission, *Annual Report to Congress of the Atomic Energy Commission for 1960,* January 1961, p. 245.

21. U.S., Atomic Energy Commission, *Annual Report to Congress of the Atomic Energy Commission for 1968,* January 1969, p. 216.

22. U.S., Atomic Energy Commission, *Annual Report for 1969,* p. 220.

23. Ibid.

24. U.S., Atomic Energy Commission, *Twenty-first Semiannual Report of the Atomic Energy Commission,* January 1957, p. 56.

25. U.S., Atomic Energy Commission, *Annual Report to Congress of the Atomic Energy Commission for 1961,* January 1962, p. 269.

26. Ibid.

27. Ibid., pp. 268–269.

28. U.S., Atomic Energy Commission, *Annual Report to Congress of the Atomic Energy Commission for 1967,* January 1968, p. 229.

29. Ibid.

30. U.S., Atomic Energy Commission, *Annual Report to Congress of the Atomic Energy Commission for 1965,* January 1966, p. 288.

31. U.S., Atomic Energy Commission, *Annual Report to Congress of the Atomic Energy Commission for 1966,* January 1967, p. 329.

32. U.S., Atomic Energy Commission, *Annual Report for 1965,* p. 285.

33. Ibid., p. 284.

34. U.S., Atomic Energy Commission, *Annual Report for 1966,* p. 330.

35. Metropolitan Edison Co., *Three Mile Island Nuclear Generating Station,* brochure, National Archives, Record Group 220, 3MI-6190068.

36. Metropolitan Edison Co., "TMI Observation Center," Radio Script no. 672-10, National Archives, Record Group 220, 3MI-6190068.

37. Metropolitan Edison Co., *Three Mile Island Nuclear Generating Station.*

38. Metropolitan Edison Co., "Science Conference," Radio Script no. 272-8, National Archives, Record Group 220, 3MI-6190068.

39. Metropolitan Edison Co., "Atomic Energy Merit Badge Clinic," Radio Script no. 1-69, National Archives, Record Group 220, 3MI-6190068.

40. Hal Stroube, "Public Acceptance of Nuclear Power," in *Nuclear Power Reactor Siting,* R. C. Ashley, ed., CONF-650201, February 1965, p. 250.

41. Ibid.

42. Ibid., p. 249.

43. Ibid.

44. Gene Pokorney of Cambridge Reports, Inc. (a polling firm), memorandum to The Electric Power Industry, 18 December 1975, p. 2.

45. Ibid., p. 5.

46. Ibid.

47. Ibid., p. 6.

48. Ibid., p. 7.

49. Ibid.

50. Ibid., p. 8.

51. Ibid., p. 9.

52. Magazine advertisement placed by Boston Edison, Eastern Utilities Association, Public Service Company of New Hampshire, and New England Gas and Electric.

53. Gordon C. Hurlbert, "The Anger of Decent Men," *Public Utilities Fortnightly,* vol. 103, no. 6, 15 March 1979, p. 19.

54. G. B. Keyes and C. G. Poncelet, Energy Action Office, Westinghouse Electric Corporation, "Addressing the Nuclear Controversy on University Campuses," proceedings of an international conference in Vienna on *Nuclear Power and Its Fuel Cycle,* vol. 7: *Nuclear Power and Public Opinion, and Safeguards* (Vienna: International Atomic Energy Agency, 1977), p. 179.

55. Atomic Industrial Forum, "Public Acceptance of Nuclear Power," *AIF Nuclear Power Information Fact Sheet,* Washington, D.C., May 1979.

56. Committee for Energy Awareness, *To Those Interested in Supporting Citizen Action* (Washington, D.C.: summer 1979), chap. 2, p. 2.

57. Ibid., chap. 2, p. 4.

58. Kenneth D. Kearns, "Citizen Action—A Key in the Nuclear Controversy," printed in *To Those Interested in Supporting Citizen Action,* pp. 10–11.

PART III: RADIATION: MANAGING THE POLITICAL FALLOUT

1. U.S., House, Subcommittee on Oversight and Investigations of the Committee on Interstate and Foreign Commerce, *"The Forgotten Guinea Pigs"—A Report on Health Effects of Low-Level Radiation Sustained As a Result of the Nuclear Weapons Testing Program Conducted by the United States Government,* 96th Congress, 2d sess., committee print 96-IFC 53, August 1980 (hereinafter referred to as *The Forgotten Guinea Pigs*), p. III.

CHAPTER 7: HOTTER THAN A $2 PISTOL

2. Ibid., p. 37.

3. U.S., House, Subcommittee on Oversight and Investigations of the Committee on Interstate and Foreign Commerce, and Senate, Health and Scientific Research Subcommittee of the Labor and Human Resources Committee, and Senate, Committee on the Judiciary, *Health Effects of Low-Level Radiation,* 96th Congress, 1st sess., 19 April 1979, serial no. 96-41 (hereinafter, *Health Effects of Low-Level Radiation*), vol. 1, p. 100.

4. Ibid., vol. 1, pp. 70–71.

5. Ibid., vol. 1, p. 100.

6. U.S., Senate, Subcommittee on Energy, Nuclear Proliferation, and Federal Services of the Committee on Governmental Affairs, *Radiation Protection,* 96th Congress, 1st sess., 6–7 March 1979, (hereinafter, *Radiation Protection*), part I, p. 2.

7. U.S., House, Subcommittee on Health and the Environment of the Committee on Interstate and Foreign Commerce, *Effect of Radiation on Human Health—Health Effects of Ionizing Radiation,* 95th Congress, 2d sess., 24–26 January, 8, 9, 14 and 28 February 1978, serial no. 95-179 (hereinafter, *Effects of Radiation on Human Health*), vol. 1, p. 146.

8. *The Forgotten Guinea Pigs,* p. 37.

9. *Radiation Protection,* part I, p. 2.

10. Ibid., p. 1.

11. U.S., Atomic Energy Commission, *Selection of a Continental Atomic Test Site,* a report approved at meeting 504, 4 December 1950, p. 4 (classification cancelled with deletions, 29 November 1978).

12. Frederick Reines, *Discussion of Radiological Hazards Associated with a Continental Test Site for Atomic Bombs,* Los Alamos Scientific Laboratory of the University of California (based on notes of meetings held at Los Alamos, 1 August 1950), LAMS-1173, 1 September 1950, p. 23 (classification cancelled with deletions, 29 January 1979).

13. *Health Effects of Low-Level Radiation,* vol. 1, p. 16.

14. U.S., Congress, Joint Committee on Atomic Energy, *The Nature of Radioactive Fallout and Its Effects on Man,* hearings before the Special Subcommittee on Radiation, 85th Congress, 1st sess., 27–29 May and 3 June 1957 (hereinafter, *The Nature of Radioactive Fallout*), part I, Atomic Energy Commission, "Report of Off-Site Radiological Safety Activities—Operation Teapot, Nevada Test Site, Spring, 1955," p. 346.

15. Ibid.

16. Ibid., p. 347.

17. *Health Effects of Low-Level Radiation,* "Never lost a fallout lawsuit, AEC said," vol. 1, p. 419, from the *Deseret News.*

18. Ibid., vol. 1, p. 221.

19. Ibid.

20. *The Nature of Radioactive Fallout,* part I, Atomic Energy Commission, "Report of Off-Site Radiological Safety Activities—Operation Teapot, Nevada Test Site, Spring, 1955," p. 348.

21. Paul Jacobs, "Clouds from Nevada," *The Reporter,* vol. 16, no. 10, 16 May 1957, p. 23.

22. Ibid.

23. U.S., Atomic Energy Commission, *27 Questions and Answers about Radiation and Radiation Protection,* 1951.

24. U.S., Atomic Energy Commission, *Atomic Tests in Nevada,* 1957, pp. 33–34.

25. Ibid., p. 23.

26. Ibid., p. 29.

27. *New York Times,* 9 June 1957, X, p. 43.

28. *Health Effects of Low-Level Radiation,* vol. 1, Atomic Energy Commission, meeting 1062, 23 February 1955, pp. 179–181 (classification cancelled 14 April 1979).

29. U.S., House, Subcommittee on Oversight and Investigations of the Committee on Interstate and Foreign Commerce, *Low-Level Radiation Effects on Health,* 96th Congress, 1st sess., 23 April, 24 May, and 1 August 1979, serial no. 96-129 (hereinafter, *Low-Level Radiation Effects on Health*), appendix C, Harold Knapp, "Sheep Deaths in Utah and Nevada Following the 1953 Nuclear Tests," 1 August 1979, with additions and revisions 9 May 1980, p. 540.

30. *Health Effects of Low-Level Radiation,* vol. 1, p. 350.

31. Telephone interview with Harold Knapp, 16 April 1980.

32. *Health Effects of Low-Level Radiation,* vol. 1, p. 256.

33. *Low-Level Radiation Effects on Health,* p. 175.

34. *Health Effects of Low-Level Radiation,* vol. 1, p. 364.

35. *The Forgotten Guinea Pigs,* p. 4.

36. *Health Effects of Low-Level Radiation,* vol. 1, p. 231.

37. *Low-Level Radiation Effects on Health,* letter from Stephen L. Brower to Utah Governor Scott M. Matheson, 14 February 1979, p. 694.

38. Ibid.

39. Ibid., letter from Stephen L. Brower to Mike Zimmerman, Utah Governor's Office, 16 February 1979, p. 696.

40. Ibid., letter from Stephen L. Brower to Utah Governor Scott M. Matheson, 14 February 1979, p. 694.

41. *Health Effects of Low-Level Radiation,* vol. 1, p. 233.

42. Ibid., vol. 1, p. 227.

43. Ibid., vol. 1, p. 230.

44. Ibid., vol. 1, "Never lost a fallout lawsuit, AEC said," p. 419, from the *Deseret News.*

45. *Low-Level Radiation Effects on Health,* letter from Stephen L. Brower to Utah Governor Scott M. Matheson, 14 February 1979, p. 694.

46. *Health Effects of Low-Level Radiation,* vol. 2, Atomic Energy Commission, "Sheep Losses Adjacent to the Nevada Proving Grounds," AEC 604/3, 4 November 1953, draft, p. 1685.

47. Ibid., Atomic Energy Commission, "Report on Sheep Losses Adjacent to the Nevada Proving Grounds," 6 January 1954, p. 1792.

48. *Health Effects of Low-Level Radiation,* vol. 1, p. 99.

49. Ibid., "Reporter's Transcript of Portions of Closing Argument of Attorney Dan S. Bushnell and Comments of the Court," David C. Bulloch, et al., plaintiffs, vs. The United States of America, defendant, Salt Lake City, Utah, 1 and 2 October 1956, case no. C-19-55, vol. 1, pp. 985–989.

50. Ibid., p. 1028.

51. Ibid., vol. 1, p. 1029.

52. Ibid., "Reporter's Transcript of Findings of the Court," David C. Bulloch et al. vs. The United States of America, 2 October 1956, vol. 1, pp. 960–961.

53. *Low-Level Radiation Effects on Health,* appendix C, "Sheep Deaths in Utah and Nevada following the 1953 Nuclear Tests, by Harold A. Knapp and Additional Comments and Responses on the Report," letter from Harold A. Knapp to Congressman Bob Eckhardt, 6 May 1980, p. 1171.

54. *Health Effects of Low-Level Radiation,* vol. 2, L. K. Bustad et al., "A Comparative Study of Hanford and Utah Range Sheep," Hanford Atomic Products Operation, Richland, Washington, 30 November 1953, vol. 2, p. 1735.

55. *Low-Level Radiation Effects on Health,* appendix C, "Sheep Deaths in Utah and Nevada following the 1953 Nuclear Tests, by Harold A. Knapp and Additional Comments and Responses on the Report," letter from Harold A. Knapp to Congressman Bob Eckhardt, 6 May 1980, p. 1180.

56. Ibid., letter from Harold A. Knapp to Dr. Leo Bustad, 29 January 1980, p. 983.

57. *Boston Globe,* 1 July 1981.

58. *Health Effects of Low-Level Radiation,* vol. 1, p. 99.

59. C. W. Thomas et al., *Radioactive Fallout from Chinese Nuclear Weapons Test September 26, 1976,* Battelle Pacific Northwest Laboratories, Richland, Washington, BNWL-2164, p. 2.

60. *Low-Level Radiation Effects on Health,* appendix C, "Sheep Deaths in Utah and Nevada Following the 1953 Nuclear Tests, by Harold A. Knapp and Additional Comments and Responses on the Report," excerpts from *Operation Upshot-Knothole, Nevada Proving Grounds, March-June 1953, Radiological Safety Operation,* AFSWP Report WT-702 (REF.), containing memorandum to Off-Site Rad/Safe Officer from Frank A. Butrico, Off-Site Monitor, "Report on the Sequence of Events occurring in St. George, Utah as a Result of the Detonation of Shot IX," p. 784.

61. *Health Effects of Low-Level Radiation,* vol. 1, p. 255.

62. Ibid., vol. 1, p. 346.

63. Ibid., vol. 2, letter from Harold Knapp to Charles L. Dunham, director, AEC Division of Biology and Medicine, 27 June 1963, comments on a memorandum to N. H. Woodruff, director, Division of Operational Safety, from Gordon M. Dunning, deputy director, Division of Operational Safety, 14 June 1963, "Comments on 'Iodine-131 in Fresh Milk and Human Thyroids Following a Single Deposition of Nuclear Test Fallout,' by Dr. Harold Knapp," pp. 1997–1998.

64. Ibid., vol. 2, p. 1997.

65. Ibid., vol. 2, p. 2000.

66. Ibid., vol. 2, p. 1999.

67. Ibid., vol. 1, p. 351.

68. Telephone interview with Harold Knapp, 16 March 1980.

69. *Health Effects of Low-Level Radiation,* vol. 1, memorandum to A. R. Luedecke, AEC general manager, from Nathan H. Woodruff, director, AEC Division of Operational Safety, 11 January 1963, vol. 1, p. 103.

70. Ibid., vol. 2, letter to D. A. Ink, AEC assistant general manager, from Gordon M. Dunning, acting director, AEC Division of Operational Safety, 27 August 1965, vol. 2, p. 2251.

71. Ibid., memorandum to AEC commissioners from Dwight A. Ink, AEC assistant general manager, 9 September 1965, vol. 2, p. 2222.

72. Telephone interview with Dr. Joseph Lyon, 30 January 1980.

73. *New York Times,* 16 February 1955, p. 1.

74. *Proceedings of the Second Interdisciplinary Conference on Selected Effects of a General War,* Defense Atomic Support Agency Information and Analysis Center, General Electric, TEMPO, Santa Barbara, California, DASA 2019-2, July 1969, vol. II, p. 41.

75. Ibid., p. 42.

76. Ibid., p. 44.

77. Ibid.

78. *New York Times,* 17 March 1954, p. 1.

79. Ralph Lapp, *The Voyage of the Lucky Dragon* (New York: Harper & Brothers Publishers, 1957), p. 148.

80. Ibid., p. 90.

81. *New York Times,* 1 April 1954, p. 20.

82. Ibid.

83. Ralph Lapp, *The Voyage of the Lucky Dragon,* p. 178.

84. *New York Times,* 1 April 1954, p. 20.

85. Samuel Glasstone, ed., *The Effects of Nuclear Weapons,* prepared by the Armed Forces Special Weapons Project of the Department of Defense (Washington,

D.C.: U.S. Atomic Energy Commission, 1957), p. 487.

86. *Proceedings of the Second Interdisciplinary Conference on Selected Effects of a General War,* vol. II, p. 124.

87. *New York Times,* 16 February 1955, p. 18.

CHAPTER 8: UNDUE ANXIETIES

1. Alvin M. Weinberg, "Salvaging the Atomic Age," *The Wilson Quarterly,* vol. 3, no. 3, summer 1979, p. 88.

2. World Health Organization (WHO), *Mental Health Aspects of the Peaceful Uses of Atomic Energy,* Report of a Study Group, Technical Report Series no. 151, Geneva, 1958, p. 12.

3. *New York Times,* 20 May 1948, p. 2.

4. WHO, *Mental Health,* annex 1, "Statement of the Sub-commitee on the Peaceful Uses of Atomic Energy of the World Federation for Mental Health, Approved by the 25th Meeting of the Executive Board of the WFMH, London, 8–12 February 1957," pp. 47–48.

5. WHO, *Mental Health,* p. 6.

6. Ibid., p. 12.

7. Ibid., p. 31.

8. Ibid.

9. Ibid., p. 33.

10. Ritchie Calder, *Living With the Atom* (Chicago: The University of Chicago Press, 1962), pp. 24–25.

11. WHO, *Mental Health,* p. 24.

12. Ibid., p. 20.

13. Ibid., p. 25.

14. Ibid., p. 28.

15. Ibid.

16. Ibid., p. 35.

17. Ibid., p. 45.

18. Memorandum to John Totter, director, AEC Division of Biology and Medicine, from Leonard Sagan, AEC Medical Research Branch, 20 November 1967, p. 3.

19. U.S., House, Subcommittee on Health and the Environment of the Committee on Interstate and Foreign Commerce, *Effect of Radiation on Human Health—Health Effects of Ionizing Radiation,* 95th Congress, 2d sess., 24–26 January, 8, 9, 14 and 28 February 1978, serial no. 95-179 (hereinafter, *Effect of Radiation on Human Health*), vol. 1, pp. 495–496.

20. Ibid., pp. 524–525.

21. U.S., House, Subcommittee on Oversight and Investigations of the Committee on Interstate and Foreign Commerce, and Senate, Health and Scientific Research Subcommittee of the Labor and Human Resources Committee, and Senate, Committee on the Judiciary, *Health Effects of Low-Level Radiation,* 96th Congress, 1st sess., 19 April 1979, serial no. 96-42 (hereinafter, *Health Effects of Low-Level Radiation*), vol. 2, L. K. Bustad, et al., "A Comparative Study of Hanford and Utah Range Sheep," Hanford Atomic Products Operation, Richland, Washington, 30 November 1953, p. 1731.

22. *Effect of Radiation on Human Health,* vol. 1, attachment, "AEC Press Release read over phone to TFM," submitted with the prepared testimony of Dr. Thomas F. Mancuso, p. 559.

23. Ibid., vol. 1, letter from Dr. J. Liverman to Dr. K. Z. Morgan, 8 September 1977, quoted in the prepared testimony of Dr. Thomas Mancuso, p. 528.

24. Ibid., vol. 1, prepared statement of Dr. Thomas Mancuso, p. 529.

25. Ibid., vol. 1, prepared statement of Dr. Thomas Mancuso, p. 528.

26. Ibid., vol. 1, p. 637.

27. Ibid., vol. 1, p. 638.

28. Ibid., vol. 1, prepared statement of Dr. Thomas Mancuso, p. 543.

29. Ibid., vol. 1, p. 638.

30. Ibid., vol. 1, p. 674.

31. Ibid., vol. 1, prepared testimony of Dr. Thomas F. Mancuso, letter from Walter H. Weyzen to Ms. Pat Borchmann, 29 March 1977, p. 573.

32. Ibid., vol. 1, p. 638.

33. Ibid., vol. 1, p. 742.

34. Ibid.

35. Ibid., vol. 1, p. 743.

36. Ibid., vol. 1, p. 718.

37. Ibid., vol. 1, p. 748.

38. Ibid., vol. 1, p. 782.

39. *Health Effects of Low-Level Radiation*, vol. 2, C. C. Lushbaugh, J. F. Spalding and D. B. Hale, "Comparative Study of Experimentally Produced Beta Lesions and Skin Lesions in Utah Range Sheep," Los Alamos Scientific Laboratory, Los Alamos, New Mexico, 30 November 1953, p. 1708.

40. Letter from Representative Paul G. Rogers, chairman, and Representative Tim Lee Carter, ranking minority member, House Subcommittee on Health and the Environment of the Committee on Interstate and Foreign Commerce, to James R. Schlesinger, secretary, Department of Energy, 4 May 1978, p. 3.

41. *Effect of Radiation on Human Health*, vol. 1, p. 720.

42. Ibid., vol. 1, p. 774.

PART IV: ONE CHANCE IN FIVE BILLION

1. Daniel Ford, "Three Mile Island—Part II," *The New Yorker*, vol. LVII, no. 8, 13 April 1981, p. 82.

CHAPTER 9: HIDING THE HAZARDS

2. U.S., Congress, Joint Committee on Atomic Energy, *A Study of AEC Procedures and Organizations in the Licensing of Reactor Facilities*, 85th Congress, 1st sess., April 1957, appendix 9, "AEC letter dated October 17, 1956, to Executive Director, Joint Committee on Atomic Energy, Forwarding AEC Reports on Certain Applications for Construction Permits," p. 145.

3. Frank G. Dawson, *Nuclear Power, Development and Management of a Technology* (Seattle: University of Washington Press, 1976), p. 88.

4. U.S., Congress, JCAE, *A Study of AEC Procedures*, appendix 8, "Report of Advisory Committee on Reactor Safeguards on PRDC Application—Made Public by AEC on October 9, 1956," pp. 133–135.

5. Ibid.

6. Dawson, *Nuclear Power*, p. 192.

7. Elizabeth S. Rolph, *Nuclear Power and the Public Safety* (Lexington, Massachusetts: Lexington Books, D. C. Heath and Company, 1979), pp. 36–37.

8. U.S., Congress, JCAE, *A Study of AEC Procedures,* appendix 6, "Correspondence in Summer and Fall of 1956 Between Joint Committee and AEC Concerning Advisory Committee Report on PRDC Application," p. 118.

9. Ibid., appendix 9, p. 152.

10. Ibid.

11. Ibid., appendix 9, p. 145.

12. Ibid., appendix 9, pp. 145–146.

13. Dawson, *Nuclear Power,* p. 193.

14. Ibid., p. 194.

15. Sheldon Novick, *The Careless Atom* (New York: Dell Publishing Co., Inc., 1969), p. 161.

16. Dawson, *Nuclear Power,* p. 194.

17. Zhores Medvedev, "Two Decades of Dissidence," *New Scientist,* vol. 72, no. 1025, 4 November 1976, pp. 264–267.

18. *London Times,* 8 November 1976, p. 5.

19. *Christian Science Monitor,* 12 January 1977.

20. U.S., Central Intelligence Agency, Teletyped Information Report, 23 May 1958, *Accidental Atomic Explosion in Chelyabinskaya Oblast,* TDCS-3/356,555 (approved for release 27 September 1977).

21. U.S., Central Intelligence Agency, Information Report, 4 March 1959, *Accident at the Kasli Atomic Plant,* CS-3/389, 785 (released 25 September 1977).

22. U.S., Central Intelligence Agency, Information Report, 16 February 1961, *Miscellaneous Information on Nuclear Installations in the USSR,* CSK-3/465,141, p. 3 (approved for release, 27 September 1977).

23. Critical Mass Energy Project Press Release, "CIA Discloses Details of Soviet Nuclear Accidents," 25 November 1977, p. 5.

24. Zhores A. Medvedev, *Nuclear Disaster in the Urals* (New York: Random House, 1980), pp. 27–128.

25. *New York Times,* 24 November 1978, p. 18.

26. W. Stratton et al., "Are Portions of the Urals Really Contaminated?," *Science,* vol. 206, no. 4417, 26 October 1979, p. 423.

27. Ibid., pp. 423–424.

28. John R. Trabalka, L. Dean Eyman, and Stanley I. Auerbach, "Analysis of the 1957–1958 Soviet Nuclear Accident," *Science,* vol. 209, no. 4454, 18 July 1980, p. 345.

29. Ibid., p. 345.

30. J. R. Trabalka, L. D. Eyman, and S. I. Auerbach, *Analysis of the 1957–58 Soviet Nuclear Accident,* ORNL-5613 (Oak Ridge National Laboratory, Oak Ridge, Tennessee, 37830: December 1979), pp. 26–27.

31. Trabalka et al., "Analysis of the 1957–1958 Soviet Nuclear Accident," pp. 347–348.

32. Ibid., p. 351.

33. Steven Ebbin and Raphael Kasper, *Citizen Groups and the Nuclear Power Controversy: Uses of Scientific and Technological Information* (Cambridge, Massachusetts: The MIT Press, 1974), p. 11.

34. David E. Lilienthal, *Atomic Energy: A New Start* (New York: Harper & Row, 1980), p. 38.

35. Ebbin and Kasper, *Citizen Groups,* p. 11.

36. U.S., Atomic Energy Commission, *Theoretical Possibilities and Consequences of Major Accidents in Large Nuclear Power Plants,* WASH-740, March 1957, p. viii.

37. Daniel F. Ford, *A History of Federal Nuclear Safety Assessments: From WASH-740 Through the Reactor Safety Study* (Cambridge, Massachusetts: Union of Concerned Scientists, 1977), p. 5.

38. Ibid.

39. Ibid.

40. Ibid.

41. Ibid., p. 6.

42. Ibid.

43. Hal Stroube, "Public Acceptance of Nuclear Power," in *Nuclear Power Reactor Siting,* R. C. Ashley, ed., CONF-650201, February 1965, p. 250.

44. Ibid., p. 250.

45. *New York Times,* 10 November 1974, p. 1.

46. Hal Stroube, "Public Acceptance of Nuclear Power," p. 250.

47. Irvin C. Bupp and Jean-Claude Derian, *Light Water* (New York: Basic Books, Inc., 1978), p. 49.

48. Ian A. Forbes et al., "Cooling Water," *Environment,* vol. 14, no. 1, January/February 1972, p. 43.

49. Ebbin and Kasper, *Citizen Groups,* p. 124.

50. Daniel F. Ford and Henry W. Kendall, "Nuclear Safety," *Environment,* vol. 14, no. 7, September 1972, p. 6.

51. Ian A. Forbes et al., *Nuclear Reactor Safety: An Evaluation of New Evidence* (Cambridge, Massachusetts: Union of Concerned Scientists, July 1971).

52. Ford and Kendall, "Nuclear Safety," p. 48, note 6.

53. Robert Gillette, "Nuclear Safety (I): The Roots of Dissent," *Science,* vol. 177, no. 4051, 1 September 1972, p. 773.

54. Ibid.

55. Memorandum to the AEC's Task Force on Emergency Core Cooling from Morris Rosen and Robert J. Colmar, 1 June 1971, "Comments and Recommendations to the REG ECCS Task Force," reprinted in U.S., *Congressional Record,* 92nd Congress, 2d sess., vol. 118, 21 March 1972, p. 9303.

56. Ibid.

57. Robert Gillette, "Nuclear Reactor Safety: At the AEC the Way of the Dissenter is Hard," *Science,* vol. 176, no. 4034, 5 May 1972, p. 498.

58. Ford and Kendall, "Nuclear Safety," p. 8.

59. Ebbin and Kasper, *Citizen Groups,* p. 131.

60. Ibid.

61. Ibid.

62. Ibid., p. 132.

63. Joel Primack and Frank von Hippel, *Advice and Dissent* (New York: Basic Books, Inc., 1974), p. 222.

64. Ibid., p. 223.

65. Ibid.
66. Daniel Ford and Henry Kendall, *An Assessment of the Emergency Core Cooling Systems Rule-Making Hearings,* Friends of the Earth, San Francisco, 1974, p. 4–10.
67. Robert Gillette, "Nuclear Safety (IV): Barriers to Communication," *Science,* vol. 177, no. 4054, 22 September 1972, p. 1081.
68. Ford and Kendall, *An Assessment of ECCS,* p. 4.14.
69. U.S., Congress, Joint Committee on Atomic Energy, *Nuclear Reactor Safety, part 2: volume I, phase IIb and phase III hearings,* 93rd Congress, 2d sess., 22–24 and 28 January 1974, "The Nuclear Power Program: Selected Quotations," submitted in testimony of Ralph Nader, p. 493.
70. Rolph, *Nuclear Power and the Public Safety,* p. 93.
71. Ford and Kendall, "Nuclear Safety," p. 9.
72. Ibid., pp. 7–8.
73. Ibid., p. 8.
74. Ford and Kendall, *An Assessment of ECCS,* p. 6.5.

CHAPTER 10: THE TOO FAVORABLE LESSON

1. U.S., Nuclear Regulatory Commission, *Reactor Safety Study, An Assessment of Accident Risks in U.S. Commercial Nuclear Power Plants,* WASH-1400 (NUREG 75/014), October 1975 (hereinafter, *Reactor Safety Study*), Executive Summary, p. 3.
2. Ibid., p. 2.
3. Ibid.
4. *New York Times,* 21 August 1974, p. 14.
5. John P. Holdren, "The Nuclear Controversy and the Limitations of Decision Making by Experts," *Bulletin of the Atomic Scientists,* vol. 32, no. 3, March 1976, p. 21.
6. Daniel F. Ford, *A History of Federal Nuclear Safety Assessments: From WASH-740 Through The Reactor Safety Study* (Cambridge, Massachusetts: Union of Concerned Scientists, 1977), p. 23.
7. Ibid.
8. U.S., Nuclear Regulatory Commission, Ad Hoc Risk Assessment Review Group, *Risk Assessment Review Group Report to the U.S. Nuclear Regulatory Commission,* NUREG/CR-0400, September 1978, p. 4.
9. Paul Slovic and Baruch Fischhoff, "How Safe is Safe Enough? Determinants of Perceived and Acceptable Risk" (draft article), Decision Research, Eugene, Oregon, January 1979, pp. 22–23.
10. P. Slovic, S. Lichtenstein, and B. Fischhoff, "Images of Disaster: Perception and Acceptance of Risks from Nuclear Power," in *Energy Risk Management,* G. Goodman and W. Rowe, eds. (London: Academic Press, 1979), p. 238.
11. Dr. William Bryan testifying before Subcommittee on State Energy Policy, Committee on Planning, Land Use and Energy, California State Assembly, 1 February 1974.
12. Ibid.
13. Richard Zeckhauser, "Procedures for Valuing Lives," *Public Policy,* vol. XXIII, no. 4, fall 1975, pp. 444–445.

14. P. Slovic, B. Fischhoff, and S. Lichtenstein, "Perceived Risk," in *Societal Risk Assessment: How Safe is Safe Enough?*, R. C. Schwing and W. A. Albers, eds. (New York: Plenum Press, 1980), reprint supplied by the authors, p. 9.

15. Henry W. Kendall, study director, Richard B. Hubbard, Gregory C. Minor, eds., *The Risks of Nuclear Power Reactors, A Review of the NRC Reactor Safety Study WASH-1400 (NUREG-75/014)* (Cambridge, Massachusetts: Union of Concerned Scientists, 1977), p. 25.

16. Ford, *A History of Federal Nuclear Safety Assessments*, pp. 16–18.

17. *Reactor Safety Study*, appendix XI, response to comments 3.2.10 and 14.9.

18. Kendall et al., *The Risks of Nuclear Power Reactors*, pp. 57–61.

19. Ibid., pp. 85–97.

20. U.S., Atomic Energy Commission, "Risks to Public From Nuclear Power Plants Very Small, Study Concludes," news release, 20 August 1974.

21. Ford, *A History Of Federal Nuclear Safety Assessments*, pp. 23–27.

22. Kendall et al., *The Risks of Nuclear Power Reactors*, p. 99.

23. Ibid., p. 104.

24. *Reactor Safety Study*, Main Report, p. 7.

25. *Reactor Safety Study*, Executive Summary, p. 5.

26. AEC news release, 20 August 1974, p. 3.

27. *New York Times*, 21 August 1974, p. 14.

28. *New York Times*, 27 August 1974, p. 24.

29. *Wall Street Journal*, 22 August 1974.

30. "New Nuclear Odds," *Time*, vol. 104, 2 September 1974, p. 69.

31. U.S., Nuclear Regulatory Commission, "Final Report of Reactor Safety Study Completed," news release, 30 October 1975.

32. Kendall et al., *The Risks of Nuclear Power Reactors*.

33. U.S., NRC, *Risk Assessment Review Group Report*, pp. 61, 66.

34. Ibid., p. 55.

35. Ibid., p. 3.

36. Ibid., p. ix.

37. Ibid.

38. Ibid., p. 3.

39. Ibid., p. viii.

40. Ibid., p. x.

41. U.S., Nuclear Regulatory Commission, *NRC Statement on Risk Assessment and the Reactor Safety Study Report (WASH-1400) in Light of the Risk Assessment Review Group Report*, 18 January 1979.

42. Interview with Norman C. Rasmussen, Massachusetts Institute of Technology, Nuclear Engineering Department, Cambridge, Massachusetts, 9 November 1979.

43. U.S., Nuclear Regulatory Commission, "Condemned to Repeat It? Haste, Distraction, Rasmussen and Rogovin," remarks by NRC Commissioner Peter A. Bradford, speaking before the Risks of Generating Electricity Section of the Seventh Annual National Engineers' Week Energy Conference, Hyatt Regency Hotel, Knoxville, Tennessee, 21 February 1980.

44. Ibid.

45. Ibid.

46. Ibid.
47. Ibid.
48. Ibid.

CHAPTER 11: BUSINESS AS USUAL

1. U.S., President's Commission on the Accident at Three Mile Island, *Report of the President's Commission On The Accident at Three Mile Island, The Need for Change: The Legacy of TMI*, October 1979, p. 9.
2. Ibid., p. 8.
3. Ibid.
4. Ibid., p. 11.
5. Ibid., p. 27.
6. U.S., Nuclear Regulatory Commission, Special Inquiry Group, *Three Mile Island, A Report to the Commissioners and to the Public,* vol. 1, January 1980, p. 90.
7. Ibid., p. 91.
8. U.S., President's Commission, *Report on TMI*, p. 10.
9. Ibid.
10. Ibid.
11. Ibid., p. 50.
12. Ibid., p. 49.
13. Ibid., p. 29.
14. Ibid.
15. Ibid.
16. Ibid.
17. Ibid., p. 43.
18. Ibid., p. 29.
19. Ibid., p. 43.
20. Ibid., p. 54.
21. Ibid.
22. Ibid.
23. Ibid., p. 45.
24. Ibid., p. 38.
25. Robert D. Pollard, ed., *The Nugget File* (Cambridge, Massachusetts: Union of Concerned Scientists, 1979), pp. 8–9.
26. Ibid., p. 12.
27. Ibid., p. 56.
28. Ibid., pp. 69–70.
29. U.S., Nuclear Regulatory Commission, Special Review Group, *Recommendations Related to Brown's Ferry Fire*, PB-249-0050 NUREG-0050, February 1976, p. 2.
30. Ibid., p. 4.
31. Ibid., p. 6.

32. Ibid., p. 7.
33. *Boston Herald American,* 3 April 1980.
34. *Boston Herald American,* 10 April 1980.
35. *Los Angeles Times,* 27 January 1974.
36. U.S., Nuclear Regulatory Commission, Office of Inspection and Enforcement, *Notice of Violation: Houston Lighting and Power Company,* docket nos. 50-498 and 50-499 (30 April 1980), appendix A, p. 2.
37. Merrill Sheils with Ronald Henkoff and Mary Lord, "Another Nuclear Scandal," *Newsweek,* 26 May 1980, vol. 96, p. 76.
38. Ibid.
39. Ibid., p. 78.
40. U.S., President's Commission, *Report on TMI,* p. 24.
41. U.S., NRC, Special Inquiry Group, *TMI,* p. 98.
42. Ibid., p. 171.
43. Nuclear Oversight Committee to President Jimmy Carter, 26 September 1980, p. 2.
44. Harold M. Agnew, "Gas-cooled Nuclear Power Reactors," *Scientific American,* vol. 244, no. 6, June 1981, p. 55.
45. Ibid.

PART V: LEAKS, SPILLS, AND PROMISES

1. U.S., Congress, Joint Committee on Atomic Energy, *Industrial Radioactive Waste Disposal,* 86th Congress, 1st sess., 28–30 January, 2–3 February 1959, p. 2086.
2. U.S., Environmental Protection Agency, Office of Radiation Protection, *State of Geologic Knowledge Regarding Potential Transport of High-Level Radioactive Waste from Deep Continental Repositories, Report of an Ad Hoc Panel of Earth Scientists,* EPA/520/4-78-004, June 1978, p. 45.

CHAPTER 12: A NONEXISTENT PROBLEM

3. U.S., Congress, JCAE, *Industrial Radioactive Waste Disposal,* p. 18.
4. Ibid., p. 12.
5. Ibid.
6. Ibid., p. 16.
7. U.S., Atomic Energy Commission, *Status Report On Handling and Disposal of Radioactive Wastes in the AEC Program,* WASH-742, 1957, p. 28.
8. U.S., Congress, JCAE, *Industrial Radioactive Waste Disposal,* p. 11.
9. U.S., Nuclear Regulatory Commission, Office of Materials Safety and Safeguards, *Essays on Issues Relevant to the Regulation of Radioactive Waste Management,* NUREG-0412, May 1978, p. 2.
10. Ibid., p. 15.
11. Ibid.
12. U.S., AEC, WASH-742, pp. 38–39.
13. U.S., Department of Commerce, *Radioactive-Waste Disposal in the Ocean,* National Bureau of Standards Handbook 58, issued 25 August 1954; as cited in U.S., Congress, JCAE, *Industrial Radioactive Waste Disposal,* p. 2999.

14. Ibid., p. 3000.
15. Ronnie D. Lipschutz, *Radioactive Waste: Politics, Technology and Risk* (Cambridge, Massachusetts: Ballinger, 1980), p. 125.
16. *Boston Globe,* 5 May 1980.

CHAPTER 13: HANFORD: THE INTERIM REALITY

1. U.S., Department of Energy, Assistant Secretary for Environment, *Assessment of the Surveillance Program of the High-Level Waste Storage Tanks at Hanford,* March 1980, pp. 7–8.
2. U.S., Congress, Joint Committee on Atomic Energy, *Industrial Radioactive Waste Disposal,* 86th Congress, 1st sess., 28–30 January, 2–3 February 1959, p. 165.
3. Ibid., p. 166.
4. Ibid., p. 165.
5. Ibid., p. 166.
6. U.S., Energy Research and Development Administration, *Waste Management Operations, Hanford Reservation, Richland, Washington, Final Environmental Statement,* ERDA-1538 UC-70, December 1975, vol. II, p. II.1-C-83.
7. Robert Gillette, "Radiation Spill at Hanford: The Anatomy of an Accident," *Science,* vol. 181, no. 4101, 24 August 1973, p. 730.
8. Ibid.
9. Ibid.
10. Ibid.
11. U.S., ERDA, *Waste Management Operations, Hanford Reservation,* vol. II, p. II.1-C-83.
12. Gillette, "Radiation Spill at Hanford," p. 728.
13. Ibid.
14. Ibid., p. 729.
15. Ibid.
16. U.S., ERDA, *Waste Management Operations, Hanford Reservation,* vol. I, p. III, 2-2.
17. U.S., Department of Energy, Office of Inspector General, *Report on Alleged Cover-ups of Leaks of Radioactive Materials at Hanford,* report no. IGV-79-22-2-231, 22 January 1980, p. 8.
18. Ibid., pp. 8–9.
19. Ibid., p. 8.
20. Ibid., pp. 18–19.
21. Ibid., appendix B, "Comments on Draft Report on Investigation of Stallos Allegations," p. 2.
22. Stephen Stallos to D. J. Cockeram, "Resignation," Rockwell International internal letter, 5 December 1978, p. 2.
23. Ibid., p. 3.
24. U.S., DOE, *Report on Alleged Cover-ups,* p. 24.
25. Ibid., p. 26.
26. Ibid., appendix A, "Attachment to Memo from Fremling to Mansfield, 21 January 1980," p. 7.

27. Ibid., p. 10.
28. Ibid.
29. Ibid., pp. 10–11.
30. Ibid., p. 11.
31. Statement of Paul Fritch, assistant general manager for Rockwell Hanford Operations, at a news conference announcing the release of *Review of Classification of Nine Hanford Single-Shell "Questionable Integrity" Tanks*, RHO-CD-896, January 1980.
32. U.S., DOE, *Report on Alleged Cover-ups*, p. 12.
33. Ibid.
34. Ibid.
35. Ibid., p. 13.
36. U.S., House, Subcommittee on Environment, Energy and Natural Resources of the Committee on Government Operations, *High Level Nuclear Waste*, 95th Congress, 1st sess., 18 June 1977, p. 94.
37. U.S., DOE, *Report on Alleged Cover-ups*, p. 13.
38. Rockwell International, Rockwell Hanford Operations, report prepared for the U.S. Department of Energy, *Review of Classification of Nine Hanford Single-Shell "Questionable Integrity" Tanks*, RHO-CD-896, January 1980, p. 4.
39. Fritch, news conference.
40. U.S., DOE, *Report on Alleged Cover-ups*, pp. 16–17.
41. Interview with John Deichman, program director for waste management, Rockwell Hanford Operations, Hanford, Washington, 22 April 1980.
42. U.S., DOE, *Report on Alleged Cover-ups*, p. 25.
43. U.S., ERDA, *Waste Management Operations, Hanford Reservation*, vol. I, pp. II, 1-44, II.1-45.
44. G. Kligfield to E. Tourigny, memorandum of U.S. Nuclear Regulatory Commission, Washington, D.C., 8 January 1979.
45. U.S., Nuclear Regulatory Commission, Office of Nuclear Material Safety and Safeguards, *Regulation of Federal Radioactive Waste Activities: Report to Congress on Extending the Nuclear Regulatory Commission's Licensing or Regulatory Authority to Federal Radioactive Waste Storage and Disposal Activities*, NUREG-05-0527, September 1979, appendix A, p. A-8.
46. U.S., ERDA, *Waste Management Operations, Hanford Reservation*, vol. II, p. II.1-C-54.
47. Tom Bauman to Steve Hilgartner, Department of Energy, Richland Operations Office, P.O. Box 550, Richland, Washington 99352, 27 June 1980.
48. Interview with Frank Standerfer, assistant manager for technical operations, Department of Energy, Richland Operations Office, Richland, Washington, 21 April 1980.
49. U.S., ERDA, *Waste Management Operations, Hanford Reservation*, vol. II, p. II.1-C-81.
50. Ibid., vol. I, p. III.1-26.
51. Ibid., vol. I, p. III.2-23.
52. Ibid., vol. II, p. II.1-C-81.
53. Ibid., vol. I, p. III.1-26.
54. Ibid., vol. II, p. II.1-C-81.

55. Ibid., vol. I, p. III.2-23.

56. Deichman interview.

57. U.S., Energy Research and Development Administration, *Alternatives for Long-Term Management of Defense High-Level Radioactive Waste,* Hanford Reservation, Richland, Washington, ERDA 77-44 UC-70, September 1977, p. 2-1.

58. Ibid., pp. 2-16–2-17.

59. Deichman interview.

60. U.S., Department of Energy, Rockwell International, Rockwell Hanford Operations, *Radioactive Waste Management at Hanford,* August 1979, p. 21.

61. Deichman interview.

CHAPTER 14: THE PERMANENT NONSOLUTION

1. State of California, Nuclear Fuel Cycle Committee, California Energy Resources Conservation and Development Commission, *Status of Nuclear Fuel Reprocessing, Spent Fuel Storage and High-Level Waste Disposal,* draft report, 11 January 1978, p. 181.

2. U.S., The President, Office of the White House Press Secretary, The White House Fact Sheet, *The President's Program on Radioactive Waste Management,* 12 February 1980.

3. State of California, *Status of Nuclear Fuel Reprocessing,* p. 181.

4. U.S., Congress, Joint Committee on Atomic Energy, *Industrial Radioactive Waste Disposal,* 86th Congress, 1st sess., 28–30 January, 2–3 February 1959, p. 2086.

5. Ibid., p. 2586.

6. U.S., Nuclear Regulatory Commission, Office of Materials Safety and Safeguards, *Essays on Issues Relevant to the Regulation of Radioactive Waste Management,* NUREG-0412, May 1978, p. 4.

7. Ibid.

8. Ronnie D. Lipschutz, *Radioactive Waste: Politics, Technology and Risk* (Cambridge, Massachusetts: Ballinger Publishing Company, 1980), p. 118.

9. U.S., NRC, *Essays on Issues,* p. 5.

10. U.S., Congress, Joint Committee on Atomic Energy, *AEC Authorizing Legislation, Fiscal Year 1972,* 92nd Congress, 1st sess., 9, 16–17 March 1971, part 3, appendix 16, "AEC Fact Sheets on National Radioactive Waste Repository, Lyons, Kansas," p. 1909.

11. Ibid.

12. Ibid., p. 1908.

13. Ibid., appendix 18, "AEC Press Releases on Selection of Lyons, Kansas, Site for Waste Repository and Transportation of Radioactive Waste," "AEC Tentatively Selects Kansas Site For Storage of Radioactive Wastes in Salt Mine," p. 1983.

14. William W. Hambleton, "The Unsolved Problem of Nuclear Wastes," *Technology Review,* vol. 14, no. 5, March/April 1972, p. 18.

15. Ibid.

16. Ibid.

17. Ibid.

18. Ibid.

19. U.S., NRC, *Essays on Issues*, p. 5.

20. U.S., Environmental Protection Agency, Office of Radiation Protection, *State of Geological Knowledge Regarding Potential Transport of High-Level Radioactive Waste from Deep Continental Repositories, Report of an Ad Hoc Panel of Earth Scientists*, EPA/520/4-78-004, June 1978, pp. 6–7.

21. Ibid., p. 10.

22. Ibid., p. 11.

23. Ibid., p. 45.

24. U.S., Department of the Interior, Geological Survey, *Geological Disposal of High-Level Radioactive Wastes—Earth Science Perspectives*, Geological Survey Circular 779, 1978, p. 3.

25. Ibid., pp. 12–13.

26. G. J. McCarthy et al., "Interactions Between Nuclear Waste and Surrounding Rocks," *Nature*, vol. 273, no. 5659, 18 May 1978, p. 216.

27. *Boston Globe*, 7 January 1979.

28. U.S., Senate, Subcommittee on Nuclear Regulation of the Committee on Environment and Public Works, *Nuclear Waste Management*, 95th Congress, 2d sess., 22 March, 4–5 April, 14 and 20 June 1978, serial 95-2, "Statement of Emilio E. Varanini, III, Commissioner, California Energy Resources Conservation and Development Commission," p. 131.

29. America's Electric Energy Companies advertisement, *New York Times*, 17 January 1980.

PART VI: MATERIAL UNACCOUNTED FOR

1. Victor Gilinsky, Commissioner, U.S. Nuclear Regulatory Commission, "Nuclear Reactors and Nuclear Bombs," speech before the League of Women Voters Education Fund, Silver Spring, Maryland, 17 November 1980.

2. *New York Times*, 14 July 1981, p. A6.

3. *New York Times*, 23 March 1978, p. A18.

4. John McPhee, *The Curve of Binding Energy* (New York: Farrar, Straus and Giroux, 1973), p. 138.

CHAPTER 15: NO EVIDENCE

5. *New York Times*, 6 August 1976, p. A14.

6. *New York Times*, 4 November 1979, p. 27.

7. Constance Holden, "NRC Shuts Down Submarine Fuel Plant," *Science*, vol. 206, no. 4414, 5 October 1979, p. 30.

8. Ibid.

9. Ibid.

10. Ibid.

11. *New York Times*, 12 December 1979, p. 1.

12. Paula DePerna, "Nuclear Debit," *The Nation*, vol. 230, no. 13, 5 April 1980, p. 389.

13. *New York Times*, 17 January 1980, p. B6.

14. "Material Losses at Erwin Less Than Half of Original Estimate," *Nucleonics Week*, 23 October 1980, p. 7.

15. U.S., Nuclear Regulatory Commission, "Amendment to Provide Exception From Procedural Rules for Adjudications Involving Conduct of Military or Foreign Affairs Functions," Rules and Regulations, 10 CFR, part 2, *Federal Register*, vol. 45, no. 130, 3 July 1980, p. 45254.

16. Ibid., p. 45255.

17. *Washington Post*, 12 October 1979.

18. Peter Bradford to Rory O'Connor, Washington, D.C.

19. *New York Times*, 6 November 1977, p. 3.

20. *New York Times*, 4 July 1977, p. 15.

21. *Boston Globe*, 6 November 1977.

22. Ibid.

23. *New York Times*, 6 November 1977, p. 3.

24. *New York Times*, 29 April 1979, p. 15.

25. Peter Bradford to Rory O'Connor, Washington, D.C.

26. *New York Times*, 12 June 1979, p. B9.

27. U.S., General Accounting Office, Report to the Congress by the Comptroller General, *Nuclear Fuel Reprocessing and the Problems of Safeguarding Against The Spread of Nuclear Weapons*, EMD-80-38, 18 March 1980, p. 13.

28. Ibid., p. ii.

CHAPTER 16: THE GAME THEORY OF SAFEGUARDS

1. *Washington Star*, 6 April 1980.

2. Interview with William H. Chambers, deputy associate director for safeguards and security, Los Alamos Scientific Laboratory, Los Alamos, New Mexico, 25 April 1980.

3. R. Jeffrey Smith, "Reprocessing Plans May Pose Weapons Threat," *Science*, vol. 209, no. 4453, 11 July 1980, p. 251.

4. Ibid., p. 250.

5. Ibid., p. 251.

6. Alvin M. Weinberg, "Nuclear Energy at the Turning Point," in *Nuclear Power and Its Fuel Cycle*, proceedings of an international conference held by the International Atomic Energy Agency in Salzburg, Austria, 2–13 May 1977 (IAEA-CN-36/593), vol. 1, p. 769.

7. Smith, "Reprocessing," p. 251.

8. Chambers interview.

9. Ibid.

10. U.S., General Accounting Office, Report to the Congress by the Comptroller General, *Nuclear Fuel Reprocessing and the Problems of Safeguarding Against the Spread of Nuclear Weapons*, EMD-80-38, 18 March 1980, p. 35.

11. Smith, "Reprocessing," p. 252.

12. Chambers interview.

13. Ibid.

14. Ibid.

15. Ibid.

16. U.S., GAO, "Nuclear Fuel," EMD-80-38, 18 March 1980, p. 11.

17. John McPhee, *The Curve of Binding Energy* (New York: Farrar, Straus and Giroux, 1973), p. 136.
18. Ibid., p. 135.
19. Chambers interview.
20. Ibid.
21. *New York Times,* 6 August 1976, p. A14.
22. Ibid.

CHAPTER 17: A NUCLEAR-ARMED CROWD

1. Interview with Udai Singh, first secretary (press), Embassy of India, Washington, D.C., by telephone, 7 August 1981.
2. *Washington Post,* 25 February 1980.
3. U.S., General Accounting Office, Report to the Congress by the Comptroller General, *Nuclear Fuel Reprocessing and the Problems of Safeguarding Against the Spread of Nuclear Weapons,* EMD-80-38, 18 March 1980, pp. 34–35.
4. R. Jeffrey Smith, "Reprocessing Plans May Pose Weapons Threat," *Science,* vol. 209, no. 4453, 11 July 1980, p. 252.
5. *New York Times,* 8 November 1980, p. 4.
6. Smith, "Reprocessing," p. 252.
7. *New York Times,* 22 May 1974, p. 7.
8. *New York Times,* 23 May 1974, p. 6.
9. *New York Times,* 9 April 1979, p. 1.
10. Albert Wohlstetter et al., *Swords From Plowshares* (Chicago: The University of Chicago Press, 1979), p. 126.
11. Ibid., p. 14.
12. Victor Gilinsky, Commissioner, U.S. Nuclear Regulatory Commission, "Nuclear Reactors and Nuclear Bombs," speech before the League of Women Voters Education Fund, Silver Spring, Maryland, 17 November 1980.
13. Wohlstetter et al., *Swords from Plowshares,* pp. 15–16.
14. Gilinsky, "Nuclear Reactors and Nuclear Bombs."
15. Amory B. Lovins, L. Hunter Lovins, and Leonard Ross, "Nuclear Power and Nuclear Bombs," *Foreign Affairs,* vol. 58, no. 5, Summer 1980, p. 1137.

PART VII: THERE ISN'T ANY CHOICE

1. Irvin C. Bupp and Jean-Claude Derian, *Light Water: How the Nuclear Dream Dissolved* (New York: Basic Books, 1978), p. 188.
2. "Notes and Comments," *The New Yorker,* vol. XLIX, no. 42, 10 December 1973, p. 37.
3. Thomas A. Vanderslice, "A Revolution of Falling Expectations," *Public Utilities Fortnightly,* vol. 103, no. 4, 1 February 1979, p. 12.

CHAPTER 18: SOFT NUMBERS AND HARD TIMES

4. U.S., Atomic Energy Commission, Office of Planning and Analysis, *Nuclear Power Growth 1974–2000,* WASH-1139-74, February 1974, p. 2.

5. M. King Hubbert, "Energy From Fossil Fuels," *Science,* vol. 109, no. 2823, pp. 103–109.

6. Palmer Cosslett Putnam, *Energy in the Future* (New York: D. Van Nostrand Company, Inc., 1953), foreword by the United States Atomic Energy Commission, 19 March 1953, p. vii.

7. Ibid., p. 2.

8. Ibid., p. 4.

9. Ibid., p. 97.

10. Ibid., p. 77.

11. Ibid., p. 108.

12. Ibid., p. 118, note.

13. Ibid., p. 169.

14. Ibid., p. 204.

15. Ibid., p. 206.

16. Ibid., p. 209.

17. Ibid., p. 215.

18. Ibid., p. 255.

19. Alvin M. Weinberg, "Outline for an Acceptable Nuclear Future," *Energy,* vol. 3, Permagon Press, Great Britain, p. 605.

20. Peter Fortescue, "The Role of Thermal and Fast Reactors in Sustaining an Adequately Safeguarded Energy Supply," in *Nuclear Energy and Alternatives,* proceedings of the International Scientific Forum on an Acceptable Nuclear Energy Future of the World, 7–11 November 1977, Osman Kemal Kadiroglu, Arnold Perlmutter, Linda Scott, eds. (Cambridge, Massachusetts: Ballinger Publishing Co., 1978), pp. 192–194.

21. Alvin M. Weinberg, "Nuclear Energy at the Turning Point," in *Nuclear Power and Its Fuel Cycle,* proceedings of an international conference held by the International Atomic Energy Agency in Salzburg, Austria, 2–13 May 1977 (IAEA-CN-36/593), vol. 1, p. 769.

22. Bupp and Derian, *Light Water,* p. 188.

23. Ibid., p. 76.

24. Ibid., pp. 48–50, 64–65.

25. Alvin M. Weinberg and Gale Young, "The Nuclear Energy Revolution," *Proceedings of the National Academy of Sciences,* vol. 57, no. 1, 15 January 1967, pp. 1–2.

26. James A. Lane, "Rationale for Low-cost Nuclear Heat and Electricity," in *Abundant Nuclear Energy,* proceedings of a symposium, 26–28 August 1968 (U.S. Atomic Energy Commission, Division of Technical Information), p. 8.

27. Ibid., p. 6.

28. Ibid., pp. 3, 25.

29. Westinghouse Electric Corporation, *Infinite Energy,* 1967.

30. Bupp and Derian, *Light Water,* pp. 73–74.

31. James A. Lane, "Rationale for Low-cost Nuclear Heat and Electricity," pp. 11–13.

32. Bupp and Derian, *Light Water,* p. 79.

33. Ibid., p. 78.

34. Marc H. Ross and Robert H. Williams, *Our Energy: Regaining Control* (New York: McGraw-Hill, 1981), p. 49.

35. A. D. Rossin and T. A. Rieck, "Economics of Nuclear Power," *Science,* vol. 201, no. 4356, 18 August 1978, p. 587.

36. U.S., General Accounting Office, *Nuclear Power Costs and Subsidies,* 13 June 1979, EMD-79-52, p. 12.

37. U.S., Department of Energy, Energy Information Administration, "Federal Support for Nuclear Power: Reactor Design and the Fuel Cycle," *Energy Policy Study,* vol. 13, February 1981, p. 59.

38. U.S., Department of Energy, Energy Information Administration, Division of Applied Analysis, pre-publication draft, "Federal Subsidies to Nuclear Power: Reactor Design and the Fuel Cycle," Joseph Bowring, author, March 1980, pp. 71–72.

39. Christopher P. David, "Federal Tax Subsidies for Electric Utilities: An Energy Policy Perspective," *Harvard Environmental Law Review,* vol. 4, no. 2, 1980, p. 335.

40. Darrel Huff and Irving Geis, *How to Lie With Statistics* (New York: Norton, 1954).

41. Atomic Industrial Forum, INFO news release, "1978 Economic Survey Results: Nuclear Power Generation Costs Stable, Reliability Improved," 14 May 1979.

42. Charles Komanoff, *Power Propaganda: A Critique of the Atomic Industrial Forum's Nuclear and Coal Power Cost Data for 1978* (Environmental Action Foundation, Washington, D.C., March 1980), p. 3.

43. *New York Times,* 6 March 1980, p. 11.

44. Edison Electric Institute, *Nuclear Power—Answers to Your Questions* (Washington, D.C., Revised 1979), p. 21.

45. U.S., General Accounting Office, *Nuclear Power Costs and Subsidies,* p. 7.

46. David Bodansky, "Electricity Generation Choices for the Near Term," *Science,* vol. 207, no. 4432, 15 February 1980, p. 724.

47. U.S., Department of Energy, Energy Information Administration, *Annual Report to Congress,* vol. 2, DOE/EIA-0173(79)/2, table 56, p. 137.

48. U.S., General Accounting Office, *Electric Powerplant Cancellations and Delays,* EMD-81-25, 8 December 1980, pp. 6–9.

49. Nigel Smith, "Wood: An Ancient Fuel With a New Future," *Worldwatch Paper,* no. 42, January 1981.

50. U.S., Department of Energy, Office of Policy, Planning, and Analysis, *National Energy Policy Plan,* July 1981, DOE/PE-0029, pp. 10–13.

51. *Boston Globe,* 1 March 1980.

52. J. J. Taylor, "The Potential Contribution of Nuclear Power in an Energy Emergency," speech given at the Contingency Planning Colloquium, 16–18 June 1980, Stanford University, Palo Alto, California, pp. 2-2, 2-3, 2-15, 5-1.

53. Milton R. Copulos, *Closing the Nuclear Option—Scenarios for Societal Change* (Washington, D.C.: The Heritage Foundation, 1978), p. 5.

CHAPTER 19: ALTERNATIVE VISIONS

1. Amory Lovins, *Is Nuclear Power Necessary* (Friends of the Earth, Ltd. pamphlet, 1979), p. 9.

2. Vince Taylor, *Energy: The Easy Path* (Cambridge, Massachusetts: Union of Concerned Scientists, 1979), pp. 18–19.
3. Amory B. Lovins, "Energy Strategy: The Road Not Taken?", *Foreign Affairs,* vol. 55, no. 1, October 1976, pp. 65–96.
4. Solar Energy Research Institute, *A New Prosperity: Building a Sustainable Energy Future* (Andover, Massachusetts: Brick House Publishing, 1981), p. 1.
5. Ibid., pp. 1, 8.
6. Robert Stobaugh and Daniel Yergin, eds., *Energy Future: Report of the Energy Project at the Harvard Business School* (New York: Random House, 1979), p. 136.
7. Demand and Conservation Panel of the Committee on Nuclear and Alternative Energy Systems, "U.S. Energy Demand: Some Low Energy Futures," *Science,* vol. 200, no. 4338, 14 April 1978, p. 15.
8. American Institute of Architects, *Saving Energy in the Built Environment: The AIA Policy,* AIA pamphlet (Washington, D.C.: 1977).
9. Robert H. Williams and Marc H. Ross, "Drilling for Oil and Gas in our Houses," *Technology Review,* vol. 82, no. 5, March/April 1980, p. 26.
10. Daniel Yergin, ed., *The Dependence Dilemma: Gasoline Consumption and America's Security* (Cambridge, Massachusetts: Center for International Affairs, Harvard University, 1980), p. 16.
11. SERI, *A New Prosperity,* p. 3.
12. Charles Drucker, "Transportation 2000: The Policy Role," *Soft Energy Notes,* vol. 4, no. 3, June/July 1981, p. 77.
13. SERI, *A New Prosperity,* p. 3.
14. Marc H. Ross and Robert H. Williams, *Our Energy: Regaining Control* (New York: McGraw-Hill, 1981), p. 80.
15. Ibid., p. 154.
16. Vince Taylor, *The Easy Path Energy Plan,* Union of Concerned Scientists pamphlet (Cambridge, Massachusetts: September 1979), p. 5.
17. *Boston Globe,* 6 September 1978.
18. *Wall Street Journal,* 22 August 1978.
19. *New York Times,* 21 June 1979, p. D1.
20. Henry W. Kendall and Steven Nadis, eds., *Energy Strategies: Toward a Solar Future* (Cambridge, Massachusetts: Ballinger Publishing Co., 1980), p. 21.
21. U.S., Department of Energy, Assistant Secretary for the Environment, Office of Technology Impacts, *Distributed Energy Systems in California's Future: Interim Report,* vol. I, prepared under contract no. W-7405-ENG-48, March 1978, pp. vi–viii.
22. Stobaugh and Yergin, *Energy Future,* p. 188.
23. *New York Times,* 14 June 1979, p. D1.
24. R. Jeffery Smith, "Wind Power Excites Utility Interest," *Science,* vol. 207, no. 4432, 15 February 1980, pp. 739–742.
25. Lloyd Bergeson, "Sail Power for the World's Cargo Ships," *Technology Review,* vol. 81, no. 5, March/April 1979, pp. 23–24.
26. *New York Times,* 14 August 1980, p. IV-7.
27. Lovins, "Energy Strategy," p. 74.
28. Ross and Williams, *Our Energy,* p. 155.

29. Environmental Defense Fund, "Utility Penalized for Ignoring Cogeneration," *EDF Letter,* January/February 1980, p. 1.

30. *Wall Street Journal,* 17 September 1980, p. 6.

31. U.S., House, Committee on Government Operations, *Nuclear Power Costs,* 95th Congress, 1st sess., 20–22 September 1977, part 2, "Statement of Randy Revelle, Seattle City Council," p. 1317.

32. U.S., Atomic Energy Commission, *AEC Authorizing Legislation Fiscal Year 1973,* hearings before the Joint Committee on Atomic Energy on Civilian Reactor Development, AEC-AL-FY1973, 92nd Congress, 2nd sess., 22–23 February 1972, p. 1107.

33. Ibid., p. 1091.

34. Ibid., p. 1107.

35. U.S., *Congressional Record,* 97th Congress, 1st sess., vol. 127, no. 99, 26 June 1981, part II, pp. H 3859–3863.

36. Nigel Smith, "Wood: An Ancient Fuel With a New Future," *Worldwatch Paper,* no. 42, January 1981.

37. Advertisement placed by America's Electric Energy Companies, Washington, D.C., in the *New York Times Magazine,* 5 July 1981, pp. 2–3.

PART VIII: ATOMS FOR WAR

1. *Los Angeles Times,* 24 January 1980, p. 1.

2. Colin S. Gray, "The Strategic Forces Triad: End of the Road?", *Foreign Affairs,* vol. 56, no. 4, July 1978, p. 774.

CHAPTER 20: THE HALL OF MIRRORS

3. U.S., House, Subcommittee on Health and the Environment of the Committee on Interstate and Foreign Commerce, *Effect of Radiation on Human Health—Health Effects of Ionizing Radiation,* 95th Congress, 2d sess., 24–26 January, 8, 9, 14 and 28 February 1978, serial no. 95-179, vol. 1, p. 341.

4. Herman Kahn, *On Escalation: Metaphors and Scenarios* (Baltimore, Maryland: Pelican Books, 1968), p. 39 (first published 1965).

5. Herman Kahn, *Thinking about the Unthinkable* (New York: Avon Books, 1968), pp. 47–48 (first published 1962).

6. H. R. Haldeman with Joseph Dimona, *The End of Power* (New York: Times Books, 1978), p. 83.

7. Bruce M. Russell and Bruce G. Blair, *Progress in Arms Control? Readings from Scientific American* (San Francisco: W. H. Freeman, 1979), pp. 130, 114.

8. Deborah Shapley, "Technology Creep and the Arms Race: A World of Absolute Accuracy," *Science,* vol. 210, no. 4362, 29 September 1978, p. 1192.

9. *Federation of American Scientists Public Interest Report:* "Command and Control: Use it or Lose it?", vol. 33, no. 8, October 1980, p. 1.

10. Ibid., p. 2.

11. Christopher Hanson, "Doomsday Option," *The Nation,* vol. 229, no. 22, 29 December 1979, pp. 676–677.

12. *New York Times,* 18 June 1980, p. 16.

13. Ibid.

14. Ibid.
15. *New York Times,* 10 June 1980, p. A16.

CHAPTER 21: THE SHATTERER OF WORLDS

1. The Committee for the Compilation of Materials on Damage Caused by the Atomic Bombs in Hiroshima and Nagasaki, *Hiroshima and Nagasaki: The Physical, Medical, and Social Effects of the Atomic Bombings* (New York: Basic Books, 1981), pp. 113–114.

2. Yoshitreu Kosaki, comp., Kiyoko Kageyama, Charlotte Susu-mago, and Kaoru Ogura, trans., *A-Bomb: A City Tells its Story* (Hiroshima: Hiroshima Peace Culture Center, 1972), p. 4.

3. Ibid., p. 8.

4. Ibid., pp. 10–11.

5. Ibid., p. 12.

6. Ibid., p. 18.

7. U.S., House, Subcommittee on Health and the Environment of the Committee on Interstate and Foreign Commerce, *Effect of Radiation on Human Health—Health Effects of Ionizing Radiation,* 95th Congress, 2d sess., 24–26 January, 8, 9, 14 and 28 February 1978, serial no. 95-179, vol. 1, p. 342.

8. *New York Times,* 16 February 1955, p. 1.

9. *New York Times,* 1 April 1954, p. 20.

10. U.S., House, Subcommittee on Oversight and Investigations of the Committee on Interstate and Foreign Commerce, and Senate, Health and Scientific Research Subcommittee of the Labor and Human Resources Committee, and Senate, Committee on the Judiciary, *Health Effects of Low-Level Radiation,* 96th Congress, 1st sess., 19 April 1979, serial no. 96-41, vol. I, "Diary of Gordon Dean, Entry Date May 27th, 1953," p. 151.

11. Lewis Strauss, chairman of the Atomic Energy Commission, 19 July 1956, quoted in Ralph Lapp, "The 'Humanitarian' H-Bomb," *Bulletin of the Atomic Scientists,* vol. XII, no. 7, September 1956, p. 263.

12. Ibid.

13. Linus Pauling, *No More War!* (Westport, Connecticut: Greenwood Press, Publishers, 1975), pp. 150–151 (first published in 1962 by Dodd, Mead & Company, New York).

14. U.S., Congress, Joint Committee on Atomic Energy, *The Nature of Radioactive Fallout and Its Effects on Man,* 85th Congress, 1st sess., 27–29 May and 3 June 1957, part I, pp. 74–75.

15. Ibid., part I, p. 82.

16. Ibid., part II, p. 1235.

17. Ibid., part I, p. 209.

18. Ibid., part II, pp. 1199–1200.

19. William A. Shurcliff, *Bombs at Bikini, The Official Report of Operation Crossroads* (New York: William H. Wise & Co., Inc., 1947), p. 152.

20. U.S., Congress, Office of Technology Assessment, *The Effects of Nuclear War,* OTA-NS-89, May 1979, p. 17.

21. Ibid., p. 20.

22. Ibid., pp. 21–22.

23. Ibid., p. 19.
24. Ibid., p. 22.
25. William J. Broad, "Nuclear Pulse (I): Awakening to the Chaos Factor," *Science,* vol. 212, no. 4498, 29 May 1981, p. 1009.
26. William J. Broad, "Nuclear Pulse (II): Ensuring Delivery of the Doomsday Signal," *Science,* vol. 212, no. 4499, 5 June 1981, pp. 1116–1120.
27. Mel Mawrence and John Clark Kimball, *You Can Survive the Bomb* (Chicago: Quadrangle Books, 1961), pp. 1–17.
28. Richard Gerstell, *How to Survive an Atomic Bomb* (New York: Bantam Books, 1950), p. 52.
29. Ibid., p. 105.
30. Ibid., pp. 21–22.
31. Ibid., p. 69.
32. Ibid., p. 82.
33. Ibid., p. 77.
34. Ibid., p. 78.
35. *New York Times,* 30 July 1950, p. 4.
36. *New York Times,* 27 December 1951, p. 10.
37. U.S., Congress, Office of Technology Assessment, *The Effects of Nuclear War,* p. 54.
38. Ibid.
39. Ibid.
40. Ibid., p. 65.
41. Ibid., p. 76.
42. Ibid., p. 27.
43. Ibid., pp. 27, 33.
44. Ibid., p. 31.
45. Ibid., p. 33.
46. Ibid.
47. Ibid.
48. Ibid., p. 35.
49. Ibid., p. 39.
50. Ibid., pp. 94–95.
51. Ibid., p. 100.
52. Ibid., p. 99.
53. Ibid., pp. 112–114.

Index

Index of Nukespeak Words

The following index includes many of the Nukespeak words italicized in the text. This index illustrates some of the general principles of Nukespeak, including: the use of euphemisms and euphoric imagery; the suppression of information and particular words; the relentless emphasis on public relations; and the restrictions of debate to a *technically qualified* elite. We hope that this index will be helpful to readers who are interested in the patterns of usage in Nukespeak.